Andreas Gebhardt
Julia Kessler
Laura Thurn

3D Printing

Understanding Additive Manufacturing

2nd Edition

HANSER

Hanser Publishers, Munich

Hanser Publications, Cincinnati

The Authors:

Prof. Dr.-Ing. Andreas Gebhardt
Managing Director, CP – Center of Prototyping GmbH, Erkelenz/Düsseldorf, Germany
Professor at the FH Aachen University of Applied Sciences, Germany
Subject specialty: high-performance methodology for production engineering and Additive Manufacturing

Dr. Julia Kessler
Managing Director, IwF GmbH, Institute for Toolless Fabrication, Aachen, Germany
Subject specialty: high-performance methodology for production engineering and Additive Manufacturing

Laura Thurn, M. Eng.
Doctoral student at the FH Aachen University of Applied Sciences, Germany
Subject specialty: high-performance methodology for production engineering and Additive Manufacturing

Cover picture: the chair for Digital Additive Production (DAP) of the RWTH Aachen University in cooperation with the Ford Motors Company designed and manufactured an improved cooling concept of a cylinder head by the use of additive manufacturing.

Distributed in the Americas by:
Hanser Publications
6915 Valley Avenue, Cincinnati, Ohio 45244-3029, USA
Fax: (513) 527-8801
Phone: (513) 527-8977
www.hanserpublications.com

Distributed in all other countries by:
Carl Hanser Verlag
Postfach 86 04 20, 81631 Munich, Germany
Fax: +49 (89) 98 48 09
www.hanser-fachbuch.de

The use of general descriptive names, trademarks, etc., in this publication, even if the former are not especially identified, is not to be taken as a sign that such names, as understood by the Trade Marks and Merchandise Marks Act, may accordingly be used freely by anyone. While the advice and information in this book are believed to be true and accurate at the date of going to press, neither the authors nor the editors nor the publisher can accept any legal responsibility for any errors or omissions that may be made. The publisher makes no warranty, express or implied, with respect to the material contained herein.

The final determination of the suitability of any information for the use contemplated for a given application remains the sole responsibility of the user.

Cataloging-in-Publication Data is on file with the Library of Congress

© Carl Hanser Verlag, Munich 2019
Editor: Dr. Mark Smith
Production Management: Jörg Strohbach
Coverconcept: Marc Müller-Bremer, www.rebranding.de, Munich
Coverdesign: Stephan Rönigk
Typesetting: Kösel Media GmbH, Krugzell
Printed and bound by Druckerei Hubert & Co GmbH und Co KG BuchPartner, Göttingen
Printed in Germany

ISBN: 978-1-56990-702-3
E-Book ISBN: 978-1-56990-703-0

Preface

Additive manufacturing (AM), 3D printing, desktop manufacturing, and some others are identical terms for the technology of layer-based manufacturing and its application.

The different terms describe these new manufacturing processes, from which the establishment of another industrial revolution is expected. They are suitable for acceleration of product development by production of complex prototypes quickly and with improved quality. But they also allow production of final parts, independent from the size of the lot.

Thus, they indeed mark a revolution in manufacturing techniques: the change from a production technology for the manufacture of large series of identical parts to a mixed series production of different parts, even down to one-of-a-kind parts.

3D printing is applicable in all branches of industry. Anybody engaged in engineering design and production, but also in strategic product planning, should know at least the basics of AM in order to perform a qualified evaluation and selection of the best applicable technology.

This book, *3D Printing*, is a new edition of *Understanding Additive Manufacturing*, which was originally published in 2011. It has been extensively updated and expanded to reflect the major new developments in the field that have taken place since then.

Suitable for the practitioner, this book imparts a basic knowledge of the processes and thoroughly demonstrates exemplary applications. Almost all currently available machines are presented in a systematic way that also allows the classification and evaluation of future systems. The large and fast-growing variety of different machines for additive manufacturing processes is also classified.

Besides processes, also discussed are new working strategies that result from the digital, mixed production, allowing a decentralized manufacture that could thoroughly change the organization of today's production.

A glossary is provided to clarify common terms and abbreviations used in 3D printing, and so to assist a quick approach into AM.

Aachen, November 2018   *Andreas Gebhardt, Julia Kessler, and Laura Thurn*

Acknowledgments

The interdisciplinary character of additive manufacturing (AM) or 3D printing and the enormous developmental speed of AM worldwide make it almost impossible for an individual to portray this discipline completely and correctly.

Therefore, we are very thankful for the diverse assistance we have received.

Many thanks to the specialists of the center of prototyping (CP-GmbH), Erkelenz, Germany, mainly for providing the insights into practice. Special thanks go to Mrs. Besima Sümer, Mr. Christoph Schwarz, and Mr. Michael Wolf.

Thanks to the members of the "GoetheLab for Additive Manufacturing", the AM Lab of the FH Aachen University of Applied Sciences, for their useful contributions. In particular we thank Alexander Schwarz (now at IWF-GmbH), Prasanna Rajaratnam, Karim Abbas, Dawid Ziebura (now at FhG-ILT), Dr. Miranda Fateri (now at DLR), Mirjam Henkel (now at LMI-GmbH), Max Kunkel (now at Siemens), and Stefan Thümmler (now at CP-GmbH).

We are very grateful for the diverse support we received from Bob Bond in terms of conception as well as regarding interpretation. Bob was the longtime Director of the Industry Grid of the Tshwane University (TUT), Pretoria, RSA, and since the late 1980s he was an early adopter of the AM technology in South Africa. For this, he was awarded the "Big Five Award of Additive Manufacturing" in 2017.

Special thanks go to the publishing house Hanser, particularly to our editor Mrs. Monika Stüve.

Andreas Gebhardt, Julia Kessler, and Laura Thurn

About the Authors

Andreas Gebhardt, born in 1953, studied mechanical engineering at the Technical University Aachen (RWTH), Germany, with the main emphasis on engine and turbine design and construction. In 1986 he passed his doctoral exam (Dr.-Ing.) at the same university with a thesis on the "Simulation of the Transient Behavior of Conventional Power Plants". From 1986 to 1991 he was general manager of a company that specialized in engine refurbishment and the manufacturing of special engines and engine parts.

In 1991 Mr. Gebhardt moved to be general manager at the LBBZ-NRW, a service center in the German federal state of North Rhine-Westphalia for the application of laser-supported material processing, where from 1992 he started working on rapid prototyping.

In 1997 the CP-GmbH (Center of Prototyping GmbH) was founded in Erkelenz/Düsseldorf, Germany, to which Andreas Gebhardt transferred as general manager. CP-GmbH is a rapid prototyping service company and manufactures prototypes from plastics and metals as one-of-a-kind or in small series. Starting with 3D CAD via additive production units to tool fabrication, CP-GmbH has at its disposal all elements of a fully closed additive manufacturing chain.

The practical experience with CP-GmbH forms the professional backbone for the subject matter of this book.

In the summer term of 2000 Andreas Gebhardt was appointed Professor for "Additive Fabrication Technology and Rapid Prototyping" at the University of Applied Sciences in Aachen, Germany. There he managed, in the framework of the "GoetheLab for Additive Manufacturing", a group of researchers, working on laser sintering of metals (SLM process), polymer printing, 3D printing (powder-binder process), the extrusion process (FDM), and applications of various fabbers. To the GoetheLab also belongs the worldwide first "Technology Bus", a rolling laboratory in a double-decker bus, called the "FabBus".

Since the winter term 2000 Andreas Gebhardt has been guest professor at the city college of the City University of New York. In autumn 2014 he was appointed "Professor Extraordinaire" at the Tshwane University of Technology, TUT, in Pretoria, South Africa.

Since 2004 Andreas Gebhardt has been editor of the RTeJournal (*www.rtejournal. de*), an "open access peer review" online journal on rapid technology.

Dr. Julia Kessler graduated as Bachelor for Bio-Medical Technology and as Master for Product Development at the University of Applied Sciences, Aachen, Germany.

From 2012 to 2015 she worked as research associate of the research group "GoetheLab for Additive Manufacturing" of the University of Applied Sciences, Aachen. Between 2015 and 2017 she was the head of the GoetheLab team that works on additive manufacturing of metals, plastics, and ceramics. Julia Kessler worked intensively on the concept of a study course for digital dental technology and a research project for jawbone augmentation by using additive manufacturing.

In cooperation with Laura Thurn she initiated and realized the project "FabBus", a mobile 3D printing laboratory. Also, the realization of the online module "Additive Manufacturing/3D Printing" was mutually developed by Julia Kessler and Laura Thurn. Within the framework of her doctorate, Mrs. Kessler worked on the structural optimization and the additive manufacturing of bionic lattice structures made from titanium and stainless steel, as well as new areas of application for these design elements.

In October 2017 she successfully passed her doctoral exam. In 2015 Julia Kessler was appointed general manager of IwF GmbH (Institute for Toolless Fabrication), which is affiliated with the University of Applied Sciences, Aachen. IwF GmbH supports industrial partners, particularly in optimization and design of the total process chain regarding additive manufacturing. Customer-oriented research and development, practice related training, and individual consulting are among the core competencies of IwF GmbH.

Laura Thurn, M. Eng. studied at the University of Applied Sciences, Aachen, Germany and gained a bachelor degree in industrial engineering with the main emphasis on mechanical engineering and a master degree in product development.

As project engineer at the "Institute of Toolless Fabrication (IwF)" she worked on the study "Generative Fertigungsverfahren in Deutschland" (GENFER; Additive Manufacturing in Germany), which investigated the potentials and challenges, and the consequences and perspectives, of the 3D printing technology in Germany.

Laura Thurn has worked since 2014 as research associate in the research group "GoetheLab for Additive Manufacturing" at the University of Applied Sciences, Aachen. She is head of the department "Plastics for AM", which, among other things, deals with the design, development, and optimization of personal printers. The activities of her group also include metallurgical investigations regarding the workability of extrusion materials and the mechanical-technological behavior of printed parts regarding their usability as products. Mrs. Thurn has dealt intensively with issues of training and further education in the field of 3D printing and conceives courses in the field of AM for different target groups. In cooperation with Julia Kessler she developed among other things the online module "Additive Manufacturing/3D Printing". Laura Thurn is one of the initiators of the rolling 3D printing laboratory "FabBus". She was co-responsible for the realization of the idea, conceives the courses in cooperation with Julia Kessler, and both mutually organize the operation.

Contents

1 Basics of 3D Printing Technology

In this book the subject of *3D printing* is considered from the user's point of view and the industrial application of *additive manufacturing* (AM) is discussed.

Chapter 1 contains a brief overview of the technology of additive manufacturing and basic principles of the *layer manufacturing process*. It includes substantial definitions as well as a classification into equipment classes for additive manufacturing.

All terms will be linked in an overview diagram, organized as *AM application levels*, which summarizes the definitions and interdependencies of the various applications. Later they will be illustrated by typical examples of different applications. The complete overview diagram of AM application levels is represented in Figure 1.1.

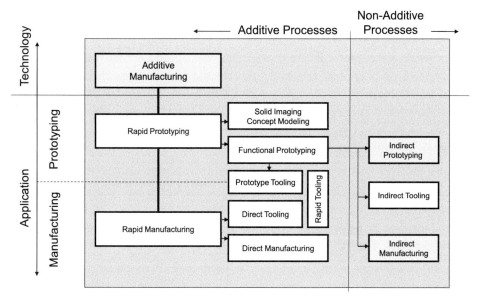

Figure 1.1 Additive manufacturing, AM; overview diagram; definition of technology and application levels

■ 1.1 Basic Terms and Definitions

1.1.1 Additive Manufacturing

Additive manufacturing (AM) is characterized as an automated process for the production of scaled, three-dimensional physical objects directly from 3D CAD data (computer-aided design). The process is based on the principle of layer manufacturing and does not require any part-dependent tools, for example a milling or a drilling device. The parts are generated by building up and connecting volume elements (voxels). Originally this process was called *rapid prototyping*, and this is often still the case today.

In combination with *subtractive production processes* like milling or turning, and *formative production processes* like casting or forging, *additive manufacturing processes* represent the third column of total production technology [1].

When the first approaches to additive manufacturing processes were released to the market in 1987, these were called rapid prototyping or generative manufacturing. Both terms are still in use and in the subsequent years many differing names were presented and frequently others were added (e.g., desktop manufacturing, freeform manufacturing, etc.). Although each of the terms is considered to be ideal from the point of view of the inventor, many of them cause confusion. This is the reason why newcomers to the field of 3D printing and additive manufacturing sometimes feel isolated.

Additive manufacturing is standardized in the U.S. by the common standards ISO/ASTM [ISO/ASTM 52900:2015].

3D printing now supersedes all other terms. The reason is that this term can be easily understood because it is more generic and of wider appeal to users. Everyone who is able to work with a text program (word processor) and to print the result as a letter by means of a 2D printer will immediately understand that, by means of a design program (part processor) and a 3D printer, a three-dimensional physical part can be produced.

The term *3D printing* is accepted worldwide as the generic term for all automated layer manufacturing processes, as is the term *3D printer*.

3D printing as a generic term should not be mistaken for the similarly named three-dimensional printing, such as the *drop on powder processes* (Section 2.1.4 "Powder-Binder Processes").

For a first overview a small selection of frequently used terms is provided to complete the group of key terms:

Table 1.1 Overview of Selected Terms and Corresponding Key Terms

Additive	3D printing Additive manufacturing, AM Additive layer manufacturing, ALM Additive digital manufacturing, DM
Layer	Layer manufacturing, layer-based manufacturing, layer-oriented manufacturing
Rapid	Rapid technology Rapid prototyping Rapid tooling Rapid manufacturing
Digital	Digital fabrication Digital mock-up
Direct	Direct manufacturing Direct tooling
3D	3D printing 3D modeling

Many possible combinations of these key terms have been created. Caution: some of these terms are protected as trademarks! Other terms are in use, which in connection with new and innovative manufacturing processes include:

- Desktop manufacturing
- On-demand manufacturing
- Freeform manufacturing

With time new terms in common use generally become the norm. Some terms are adopted and promoted by manufacturers, and some of these compete for wider acceptance.

1.1.2 The Principle of Layer-Based Processes

The terms *3D printing, additive manufacturing,* and *AM* include any possible process regarding the arrangement of material for the production of a physical part. The technical execution of AM is exclusively based on layers and therefore is called *layer-based technology* or *layered technology*. Therefore, today terms such as 3D printing, additive manufacturing, and layer-based technology are in use.

If in future new additive technologies become available, these need to be integrated into the structure of the AM terms. For example, a process called *ballistic particle manufacturing (BPM)* was introduced in the early 1990s, but vanished shortly thereafter. In this process, material was fed to the part from all directions of the build space by attaching discrete volumes (voxels) to the developing object by means of jets. This technology is considered as additive but not layer-based.

Additive Manufacturing Process Chain

Additive manufacturing processes are characterized by a process chain, which is represented in Figure 1.2. The process starts with a (virtual) three-dimensional CAD data file, which represents the part to be produced. In the engineering phase the data file will be typically generated by means of 3D CAD design (CAD), scanning, or imaging processes like computerized tomography (CT scanning).

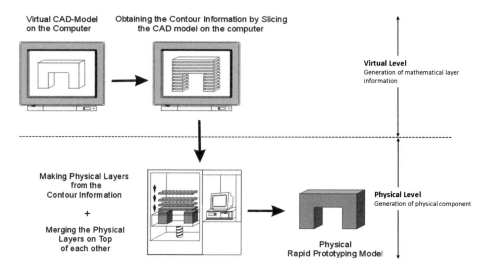

Figure 1.2 Additive manufacturing; process chain

Independent of how the 3D data set is generated, in a first step it is split by means of special software into slices or layers. As a result a data set containing contour data (x-y), thickness data (dz), and the layer number (or z-coordinate) of each layer will then be transmitted to a machine, which carries out two basic process steps for each layer, to generate the part.

In a first step each layer will be generated according to the defined contour and layer thickness. This can be performed in different ways using different physical principles. The simplest method is to cut the contour out of a foil or a sheet.

In a second step, each layer will be connected with the previous one and the new layer then forms the top layer of the growing part. Layer by layer the physical model thus grows from the bottom to the top until the part is finalized.

Depending on the uniform thickness of the layers, all parts produced by means of layer technology show a so-called *stair-step* effect, as shown in Figure 1.3 (right).

Figure 1.3 Principle of layer manufacturing: contoured layers (left); 3D-object; generated by staggered layers (right)
(Source: HASBRO/MB puzzle)

These basic steps, called process chains, are the same for more than 200 different AM machines. The machines only differ in the method of generating single layers and in how adjacent layers are connected to form the part.

Therefore, all machines which are discussed in Chapter 3 "The Additive Manufacturing Process Chain and Machines for Additive Manufacturing" are characterized by numerous common properties.

- All machines use a 3D data base, a three-dimensional virtual object, which also is called a *digital product model.*

- The manufactured parts are all generated by means of layers with uniform thickness, related to the cross-sections of the product model. Additive manufacturing thus represents basically a 2½D process.

- The build processes do not have any reciprocal (or feedback) effect regarding the product development. They are solely facsimiles of the digital product model.

- The machines therefore can be used in any stage of product development and for production.

- The build processes use process-oriented and therefore generally machine-specific materials. This provides a close linkage between machine, process, and build material.

In this context it needs to be stressed that additive manufacturing processes are production processes.

■ 1.2 Application Levels

People interested in additive manufacturing generally like to know how this new technology can be used and which novel products can be developed from it. Moreover, it is imperative to use correct and uniform terms in discussions within product development groups.

Often it is assumed that each of the different additive manufacturing processes, which are described in detail in Chapter 2 "Additive Manufacturing Processes/3D Printing", is linked to a certain application in a way such that a specific AM process can only be used for a particular application or a small number of applications. This interpretation results in interested people studying multiple processes, before they start to deal with suitable applications.

Practically, the selection of the best suitable additive manufacturing process starts with the requirements of a specific application. In the next step, certain requirements like dimensions, requested surface quality, allowable mechanical loads, temperatures, etc. lead to the selection of the suitable material and finally to the selection of a process or a machine that can fulfill these requirements to a satisfactory degree. Basically, different additive processes are suitable to address a specific requirement.

Before the various additive manufacturing processes are presented (Chapter 2 "Additive Manufacturing Processes/3D Printing"), the application fields have to be structured.

- It is necessary to distinguish between the terms *technology* and *application*. *Technology* is defined as the science of a technical process. It describes the scientific disposition.

- *Application* is understood as use of the technology for the benefit of the user and is therefore also defined as practical disposition. For a better overview, so-called *application levels* are defined. These definitions are generally accepted, although they are not yet standardized and, despite all endeavors for standardization, different terms are partly in use. As designated in Figure 1.4, AM technologies are characterized by the main application levels *rapid prototyping* and *rapid manufacturing.*

Rapid prototyping includes all applications that result in prototypes, models, or mock-ups. *Rapid manufacturing* is applied when final products or simply products are to be generated.

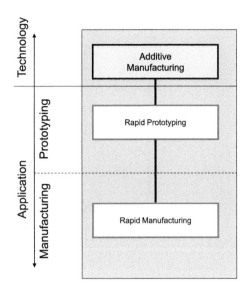

Figure 1.4 AM: Technology level and the two application levels rapid prototyping and rapid manufacturing

1.2.1 Direct Processes

All additive manufacturing processes are named *direct processes* to express that, out of the digital data model by means of an additive machine, a physical object – called a part – is directly generated. In contrast, some processes are named *indirect processes*. These processes do not apply the principle of layer manufacturing, and are consequently not considered as AM processes. Indeed, indirect processes are copy work, which normally is executed as vacuum casting using silicone molds (see Section 1.2.2 "Indirect Processes").

As for indirect processes, additively generated parts (prototypes) are used as masters; the term *indirect rapid prototyping process* was introduced, probably because it sounds more innovative. The procedure is described in Section 1.2.2 "Indirect Processes".

1.2.1.1 Rapid Prototyping

With regard to the application level rapid prototyping, two sublevels can be identified: *solid imaging* (generating a three-dimensional mock-up or a sculpture) on the one hand and *concept modeling* (generating a concept model) on the other (Figure 1.5 and Figure 1.8).

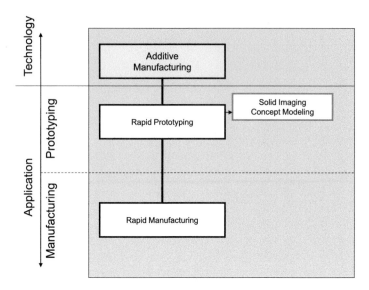

Figure 1.5 AM: application level rapid prototyping; sublevel solid imaging and concept modeling

By solid imaging or concept modeling a parts family is generated, which serves to verify a basic concept. The parts resemble a three-dimensional image or a sculpture. In most cases, they cannot be loaded physically. They merely show a special representation to judge the general appearance and the proportions. Therefore, these parts are also called *show-and-tell models*. Scaled concept models are often used to illustrate complex CAD drawings. In this context, they are also designated as *data control models* (Figure 1.6). The examination of the dimensions not only serves to check the CAD data, but is also the basis for upcoming interdisciplinary discussions, e.g. regarding packaging problems. Concerning the concept model for the roof of a convertible (cabriolet), according to Figure 1.6, it allows a balance of ideas in the involved design departments regarding the different aspects of the convertible roof – the electrical drive as well as the kinematics.

Colored models manufactured by means of 3D printing (see Section 2.1.4 "Powder-Binder Process") are estimation tools for concept development. Coloring helps in the recognition of difficult zones of a product and to shorten discussions. Figure 1.7 shows a solid image of a cut-away model of a combustion engine. Different colors of the model can be, for example, linked to the main topics of the items for discussion. In reality, the part is of course not colored.

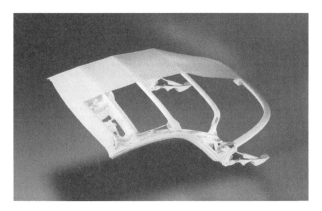

Figure 1.6 Solid image or concept model; scaled arrangement of a roof design of a
convertible; laser sintering, polyamide
(Source: CP-GmbH)

Figure 1.7 Solid image or concept model. Cut-away demonstration part of a combustion
engine; 3D printing
(Source: 3D Systems)

Functional prototyping (see Figure 1.8) is applied to examine and verify one or
multiple separate functions of the later product or to take the decision for the production, even if the model cannot be used as a final part [2].

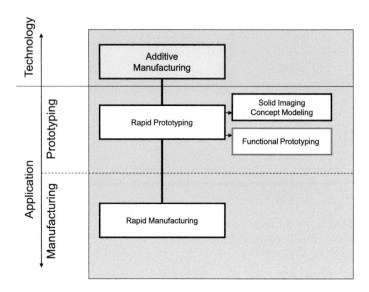

Figure 1.8 AM: application level rapid prototyping; sublevel functional prototyping

As shown in Figure 1.9, the model of an adjustable air-outlet grille for the air conditioning of a passenger car can be used to check the air distribution at a very early stage of the product development. The part was manufactured in one piece by means of laser stereolithography.

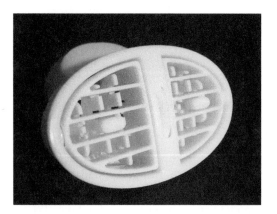

Figure 1.9 Functional prototyping: adjustable air-outlet grille for a passenger car; laser stereolithography
(Source: 3D Systems)

The stereolithography process provides a smooth surface, imitating the quality of the subsequent serial manufacture. However, this kind of manufacturing is not suitable for serial manufacture, with respect to the mechanical and especially the thermal properties of the material, as well as to the color and the final price.

To manufacture the movable parts, a connecting layer within the hinges was not cured (see Section 6.2.4 "Flexures, Hinges, and Snap-Fits"). Finally, the finished part was cleaned and specifically the uncured material was removed (mainly manually). After that, the device was ready for testing.

Figure 1.10 shows the casing of a mobile phone which was designed for the installation of a local telecommunication grid for poor communities. The mobile phone was derived from a low-cost walkie-talkie (portable two-way radio equipment). To use it as a mobile phone, the loudspeaker and microphone had to be arranged in a way that simultaneous speaking and listening was possible, and ergonomic handling was achieved. The two-piece test casing is made from ABS plastics by *fused deposition modeling* (FDM) (see Section 2.1.3 "Extrusion/Fused Layer Modeling").

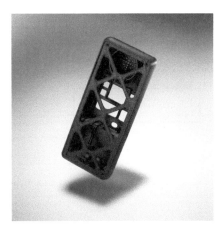

Figure 1.10 Casing for a mobile phone; extrusion process: fused deposition modeling FDM
(Source: GoetheLab, University of Applied Sciences, Aachen)

The lower part of the casing is designed to house the electronics, while the upper part serves as cover for the casing. Both parts have to fit perfectly.

The prototype of the casing was used to prove the correct fit as well as for testing of the handling. Due to the clearly visible extrusion structure and the production costs, which are too high for serial production, use as a final product has to be ruled out.

1.2.1.2 Rapid Manufacturing

The application level *rapid manufacturing* includes all processes that produce final products or deliver parts that have to be assembled afterwards to produce a product. A part generated by *additive manufacturing (AM)* will be designated as (final) product if it shows all properties and functions which have been determined during the development process of the product. If the generated part is a positive, the process is called *direct manufacturing*. In the case of a negative, e.g. a die, mold, or gauge, it is called *direct tooling*. Direct manufacturing leads to products which are generated directly by means of an AM process (Figure 1.11). A large variety of materials of all main material types (plastics, metals, and ceramics) is available (see Section 6.1.2 "Isotropic Basic Materials").

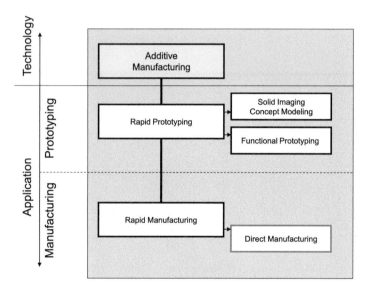

Figure 1.11 AM: application level rapid manufacturing; sublevel direct manufacturing

It is essential for the function of the part that material and manufacturing process are generating exactly those mechanical-physical properties which have been defined during the design process (generally the engineering). If this is achieved, they mimic their behavior.

Figure 1.12 shows a dental bridge, made of three elements from a CoCr alloy, which was manufactured by means of *selective laser melting, SLM*. The data file was generated by a digitized dental imprint of the patient.

The dental bridge was designed applying professional dental software (3 shape), and directly manufactured by means of SLM. After finishing and adjustment, the bridge was ready for fitting to the patient. Compared with traditional technology, the production of a directly manufactured bridge was faster with customized fit and comparable costs.

Figure 1.12 Direct manufacturing: dental bridge (three elements directly after manufacturing; without removing of support structure (left), after finishing (right); selective laser melting (SLM), CoCr alloy
(Source: GoetheLab, University of Applied Sciences, Aachen)

A hinge for the cover of an aircraft turbine (Figure 1.13, above) was redesigned for the production by means of direct manufacturing and then tested. A bionic design was applied, which led to a weight reduction of 50%. However, it could no longer be produced by traditional machining. It was finally manufactured by means of an AM metal process, by selective laser melting (SLM). The part passed the customary tests and functioned perfectly (Figure 1.13).

Figure 1.13 Direct manufacturing: hinge of a cover of an aircraft turbine (below) compared to a traditional manufactured hinge (above). Selective laser melting, SLM; stainless steel
(Source: EADS)

1.2.1.3 Rapid Tooling

Rapid tooling encompasses all additive manufacturing processes, resulting in cores, cavities, or inserts for tools, dies, and molds. In addition, two sublevels can be distinguished: *direct tooling* and *prototype tooling*.

From the technical point of view, direct tooling is equivalent to direct manufacturing, but is confined to tool inserts, dies, and molds, which are produced for quantity serial manufacture (Figure 1.14).

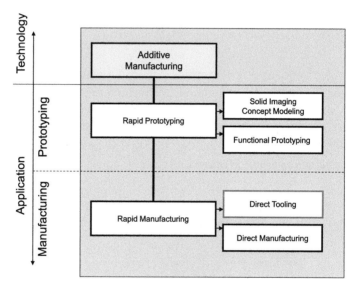

Figure 1.14 AM: application level rapid manufacturing, sublevel direct tooling

Although design of tools is the inversion of the product data set (inversion of a positive into a negative), it requires complex detail design.

Moreover, production by means of tooling requires the definition of parting areas, consideration of shrinking allowance, provision of chamfers with release angles for de-molding, and design of ejectors, etc. Tool production usually requires a metal-handling process and machines which have been designed to facilitate this process.

Direct tooling does not mean that a complete tool is manufactured. In fact, mainly cavities (tool inserts) or sliders are generated. The complete tool is made by assembling the inserts and standard components, as it is done during conventional tool manufacturing.

The layer-based technology of all additive manufacturing processes enables the generation of tool-internal hollow structures. This allows the production of mold inserts with internal conformal cooling channels (Figure 1.15, right) which follow the outlines of the tool insert below its surface.

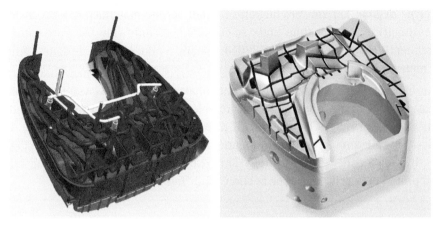

Figure 1.15 Direct tooling: mold insert with conformal cooling channels (dark) and pneumatic ejectors (white); laser sintering/laser melting (laser fusing); tool steel (Source: Concept Laser GmbH)

This method is called *conformal cooling*, as the arrangement of the cooling channels follows the outlines of the mold. Due to the forced cooling, the cycle time for a plastics injection molding machine can be reduced significantly and thus the productivity increases. Moreover, it is possible to design an integrated heat distribution system by means of the arrangement of heating and cooling channels and thus create a more efficient tool.

To generate a steel mold for the production of golf balls high precision is required. By means of *direct laser metal sintering* (DLMS, technically equivalent to SLM) a near net shape structure was manufactured additively (Figure 1.16). It is a good example of how by an additive manufacturing processes and subsequent high-precision mechanical processing, like high-speed milling and electric discharge machining (EDM), the total process can be optimized.

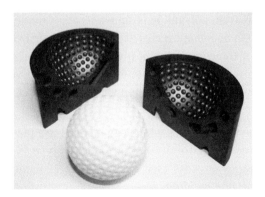

Figure 1.16 Steel mold for injection molding; direct laser melting (SLM) (Source: EOS GmbH/Agie Chamilles)

Prototype tooling: For the production of small series, manufacture of a mold in series quality is often too time- and cost-consuming. If only a few parts are required, or frequently details have to be modified, usually a pilot mold made from a substitute material meets the requirements.

Molds of this type show the quality of functional prototypes; however, they belong – at least partly – to the functional application level *direct tooling*. The corresponding application level ranges between *rapid prototyping* and *rapid manufacturing*. This sublevel is called *prototype tooling* (manufacturing of prototype tools); see Figure 1.17. It is also called *bridge tooling*. This expression is also used for secondary rapid prototyping processes (see Section 1.2.2 "Indirect Processes").

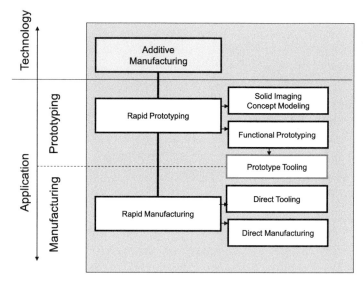

Figure 1.17 AM: application level rapid prototyping/rapid manufacturing; sublevel prototype tooling

A prototype tool made from polyamide is shown as an example in Figure 1.18. It is used for the production of a small series of a newly designed sole of a rubber boot. The soles are used for the completion of the prefabricated bootleg, without manufacturing in advance metal tools suitable for series production. Different profiles and materials for the soles can be evaluated very quickly by casting, even on small budgets.

A prototype tool made from plastics which can be used in a plastics injection molding machine is shown in Figure 1.19. The tool prototype is manufactured in a special stereolithography process (see Section 2.1.1 "Polymerization") which is named *AIM, ACES injection molding.*

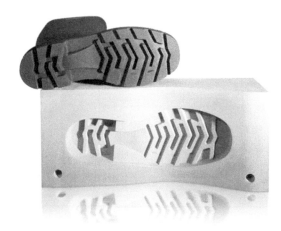

Figure 1.18 Prototype tooling: model of a sole of a rubber boot. Laser sintering; polyamide (PA)
(Source: EOS GmbH)

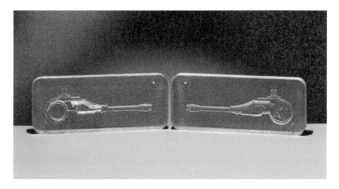

Figure 1.19 Prototype tooling: AIM injection molding; mold insert; stereolithography
(Source: 3D Systems)

ACES stands for *accurate clear epoxy solid* and is a proprietary stereolithography build style of 3D Systems, Inc. [2, 3].

Both halves of a mold are manufactured – preferably with thin wall thickness – simultaneously by means of AM stereolithography and then are backed with heat conducting material such as aluminum-filled epoxy.

AIM is suitable for injection molding of simply shaped parts with a low-volume production.

To summarize the different kinds of AM toolmaking, it is demonstrated that *rapid tooling* does not represent an autonomous application level (Figure 1.20). *Rapid tooling* encompasses all applications of additive manufacturing processes that produce dies, molds, or comparable inserts.

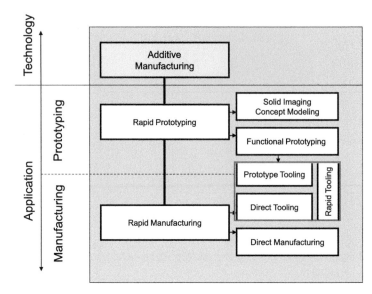

Figure 1.20 AM: rapid tooling, defined as a sub-category, integrating prototype tooling and direct tooling

1.2.2 Indirect Processes

Additive manufacturing produces a geometrically exact and scaled physical image of the virtual data set. However, this process also shows disadvantages, at least with respect to the majority of actual additive processes (see Chapter 2 "Additive Manufacturing Processes/3D Printing" regarding details).

Additive Manufacturing Processes
- Work with process- and consequently machine-dependent materials, and are restricted with regard to coloring, translucence, transparency, and elasticity.
- Do not result in significant cost reductions with increasing production volumes.
- Consequently are relatively expensive when manufacturing numerous parts, particularly in series production.

To overcome these disadvantages, additive manufactured parts can serve as master models for subsequent copying or reproduction processes. This procedure often is named the separation of properties: the geometrically exact master is quickly generated by means of an AM process, while the required volume and the part's properties (like coloring, etc.) are obtained by subsequent copying processes.

The copying or follow-up process is not a layer-based process and therefore cannot directly be assigned to *additive manufacturing*. Consequently, it is called an *indirect process*. For marketing reasons and to underline the manufacturing speed, it is called an *indirect rapid prototyping process*. The literature sometimes uses the expression *secondary rapid prototyping process* for the same reason.

1.2.2.1 Indirect Prototyping

Indirect prototyping is applied to improve the properties of an AM part in order to fulfill the requirements of the user, if an additively manufactured part does not represent this directly. If, for example, an elastic part is required and cannot be generated directly by means of an additive manufacturing process due to material restrictions, a geometrically exact rigid AM part is manufactured and used as master model for a subsequent casting process (Figure 1.21).

Possible shrinkage will be compensated for by scaling of the master model used for production.

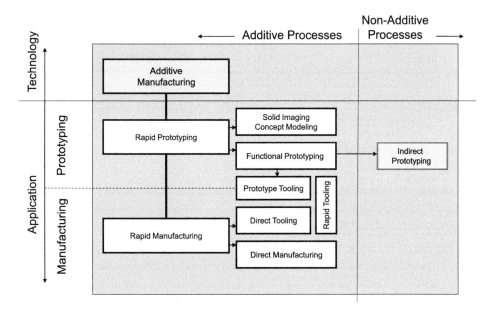

Figure 1.21 AM: indirect processes, indirect prototyping

Many different secondary processes are applied (for details see Chapter 3 "The Additive Manufacturing Process Chain and Machines for Additive Manufacturing" and [2]). The best known process is *vacuum casting* or *room temperature vulcanization, RTV*, which is also called *silicone rubber molding*. Most of the secondary processes as vacuum casting are totally or partly manual processes with long cycle times and therefore only applicable for prototyping, small series, or one-of-a-kind production.

For the "system plug" (Figure 1.22), plug housings with different colors as well as partly transparent ones are required. Based on an additively manufactured master in two parts for the housing, a silicone mold was generated. With this mold, approximately 15 different copies were made by means of vacuum casting.

Figure 1.22 Indirect prototyping: vacuum casting; system plug, master model made by means of stereolithography; mold with upper part of the plug housing (left), assembled plugs (right)
(Source: meis van shoen Design/CP-GmbH)

The cast parts served to be presented to an optional series producer in order to illustrate the new product and its qualities. The parts are prototypes, made from prototype material and thus not series products, even if they are functioning well.

The internal evaluation of the table lighter "Bruce", designed for Alessi by Stefano Giovannio in 1998, was very important for the production decision. As the table lighter was operated with a disposable lighter which was inserted in its base, while the flame was to exit at the top, easy and safe handling of the table lighter was required. Since in this stage of product development steel tools would have been too expensive, the test items were manufactured by means of vacuum casting, based on a master model, which itself was generated by means of stereolithography. Figure 1.23 shows the stereolithography master-model in front and some colored versions with and without mechanism.

Prototype parts made from soft materials, e.g. gaskets, frequently have very complex geometries. This is the case for gaskets for the installation of car mirrors, known as "mirror triangles". They have to provide multiple functions like sealing against water, fastening of the mirror and parts of the window, electrical wiring run-through, attractive appearance, and integration of adjacent extruded sealings.

Figure 1.23 Indirect prototyping: vacuum casting. Table lighter *"Bruce"*, AM master-model;
stereolithography; functional copies by vacuum casting
(Source: Alessi/Forum Omegna)

Figure 1.24 shows a sealing which was manufactured by means of vacuum casting.
The cast part is based on a scaled rigid master model generated by stereolithography.

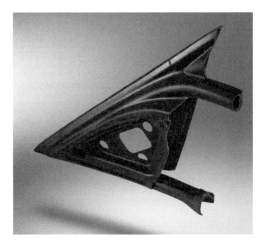

Figure 1.24 Indirect prototyping: vacuum casting; triangle shaped gasket for car mirror fixation
(Source: CP-GmbH)

1.2.2.2 Indirect Tooling

Indirect tooling is based on the same copying processes as all indirect processes (Figure 1.25). The aim is not to manufacture a final part, but a tool which forms the basis for the production of a small- or medium-scale batch production of final parts or products.

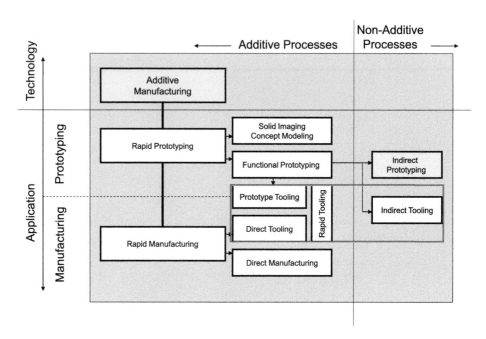

Figure 1.25 Indirect processes; indirect tooling

Compared to serial tools made from tool steel, tools made by *indirect tooling* can be manufactured faster and more economically (by casting).

Comparable to indirect manufacturing of prototypes, an additively generated master model can also be used for the indirect manufacturing of tools, minimizing mechanical working, grinding, and eroding. The requirement is to manufacture a tool for a small- or medium-volume series production based on a printed functional prototype.

In contrast to vacuum casting, the master model must be suitable for the production of a larger number of parts. The final products may consist of plastics and metals. Against this background indirect tooling can be considered as part of rapid tooling, although it is not a layer-orientated process.

For example, a mold for the manufacturing of wax cores for investment casting (lost wax casting) is shown on Figure 1.26. The mold is generated on the basis of an additively manufactured master model, by counter-casting it from polyurethane

(PUR) in an aluminum back-up frame. After removal of the master the mold is used for the production of the required number of wax cores. Due to the higher stiffness of the PUR material combined with the supporting aluminum frame a mold is formed, which produces significantly more precise wax cores, as could be received from a soft silicone mold. In comparison with mechanically milled aluminum tools this method is more economic in combination with a much shorter manufacturing time. These wax models can be used to manufacture (small) series of complex precision casts.

Figure 1.26 Indirect tooling: PUR mold, based on an AM master. Open mold half with cast wax master for investment casting; PUR mold (black); supporting frame from aluminum (Source: BeNe Aachen)

However, there are also parts which cannot be manufactured as models by means of a PUR mold, but have to be manufactured from the final series material by means of plastics injection molding, such as specific plastic parts made from flame-resistant materials. Applying injection molding requires rigid molds. To avoid steel or aluminum tools a suitable mold may also be manufactured from epoxy resin which is filled with aluminum powder to increase strength and heat resistance; these are called cast resin tools. The required master model is generated be means of stereolithography or polymer printing.

For the improvement of details and to generate sharp edges, which are difficult to cast, mechanically machined inserts can be integrated into the epoxy master by casting.

Cast resin tools usually are not cooled and, instead of sliders, are equipped with manually handled inserts. Normally, the tools are demolded manually, with the disadvantage of long cycle times.

Figure 1.27 shows both cast resin mold-halves before installation into the mold frame. The master models (bright) manufactured by means of stereolithography and a set of parts generated by injection molding (black) can be seen.

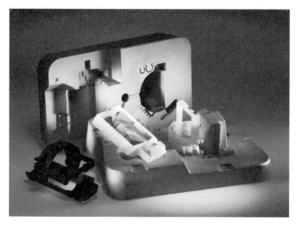

Figure 1.27 Indirect tooling: rigid two-parts mold made from aluminum-filled epoxy for injection molding of identical series parts: injection molded parts (black, left); stereolithography master model (bright, center) (Source: Elprotec GmbH)

A small series of parts made from HDPE was tested in the engine compartment of a passenger car.

1.2.2.3 Indirect Manufacturing

Indirect manufacturing is also based on additively manufactured master models. The aim is to generate one-offs or small series with properties that equal traditionally manufactured products (non-additively manufactured products).

Consequently, indirect manufacturing is part of the application level *manufacturing* (Figure 1.28).

As example for indirect manufacturing the scaled engine block of a six-cylinder engine made from aluminum is shown in Figure 1.29. It was manufactured by evaporative pattern casting (also called full-mold casting) as one-of-a-kind production.

The "lost master model" (or "lost foam model") was generated by means of laser sintering from polystyrene. For each casting procedure, a new master model is required.

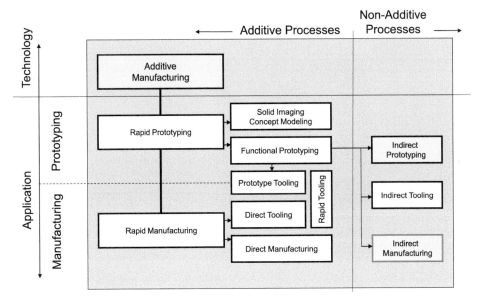

Figure 1.28 AM: application level rapid manufacturing sublevel: indirect manufacturing

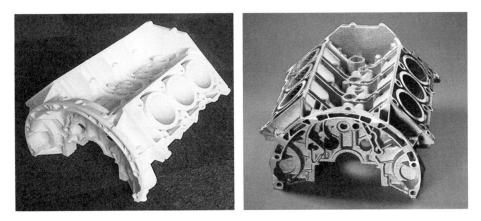

Figure 1.29 Indirect manufacturing: motor block of a six-cylinder engine. AM master–model (left), laser sintering, polystyrene; aluminum cast, one-of-a-kind part (right) (Source: Grunewald)

As a result, a series of identical motor blocks was produced. The process can be applied for the optimization and evaluation of the engine design in test runs, long before series products are available. This method is used for the manufacturing of small series, such as racing engine blocks. Whether this can be considered as an appropriate manufacturing method is not a technical but a commercial issue.

The same process was applied for the production of an air-inlet distributor, shown in Figure 1.30. It was made from aluminum by investment casting. As shown in Figure 1.29, the master model was generated by means of laser sintering from polystyrene. The picture to the left shows the part after the preparatory surface treatment with wax, and the picture to the right shows the cast part.

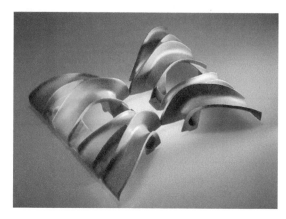

Figure 1.30 Indirect manufacturing: air-inlet distributor, AM master-model, polystyrene; laser sintering after finishing (left); aluminum cast, one-of-a-kind production (right) (Source: CP-GmbH)

Another variation of these processes is given in Figure 1.31. It demonstrates that the 3D printing process also allows the production of precise casts by use of PMMA plastics. The picture shows a gearbox as lost mold, as well as the cast part, for a racing car.

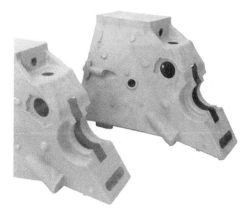

Figure 1.31 Indirect manufacturing: gearbox for a racing car; AM master-model, made from PMMA by means of 3D printing (left); aluminum casting, one-of-a-kind production (right)
(Source: Voxeljet)

■ 1.3 Classification of Machines for Additive Manufacturing

A great variety of machines for additive manufacturing is available on the market. The different machines can be subdivided into classes, which are not strictly linked to the applied additive manufacturing process. For example, extrusion machines can be part of *personal printers, professional printers, production printers,* or *industrial printers* (Table 1.2).

Generally, a machine for layer-oriented additive manufacturing processes is called *fabricator* particularly if it is producing final parts. If only prototypes can be generated, it is mostly called a *prototyper.*

The trend is to name all types of machines which are working according to the layer-oriented additive manufacturing process a printer or 3D printer, often using a prefix such as personal, professional, production, or industrial.

1.3.1 Generic Terms for AM Machines

Umbrella terms have been developed that attribute all machines for additive manufacturing available on the market to four categories or classes: personal printer, professional printer, production printer, and industrial printer (see Table 1.2 and Chapter 3 "The Additive Manufacturing Process Chain and Machines for Additive Manufacturing").

Generally, the term *personal printer* is used to characterize especially small, simple, and economical machines. Personal printers can be subdivided into fabbers and desktop printers. *Fabbers* are mainly privately used personal 3D printers, which in most cases are built or assembled by the users themselves. The expression "fabber" is the shortened form of *fabricator* and characterizes the philosophy behind it, i.e., that by means of such machines everybody can manufacture (almost) everything. *Desktop printers* are 3D printers used predominantly in the professional or semi-professional field. Although these machines are in the lower price bracket, they are not for do-it-yourself assembly, but rather simple printers with sufficient high accuracy which can be operated by people who can handle a computer after a short training. Productivity is not at the forefront (see Section 3.2.1 "Personal Printers").

Professional printers, professional 3D printers, or office printers are compact, simple-to-operate machines with low maintenance requirements, which can be operated in a bureau or a workshop, providing a higher level regarding quality of the apparatus and its elements. They are equipped with material changing systems, part discharge to avoid contamination of operators and rooms, and suitable (partly manual) external systems to remove support structures. Training for the handling of the machines is provided by a short supplier instruction.

Table 1.2 Generic Terms for AM Machines and Their Assignment to Application Levels

Term			
Personal printer	Professional printer	Production printer	Industrial printer
Also called:			
Personal fabricator/ Fabber Desktop printer	Office printer/ Printer	Shop floor printer/ Printer	Flexible manufacturing systems; Flexible AM systems (FAMS)
3D printer/ Professional 3D printer	Professional 3D printer	Production 3D printer	Industrial print systems
Application			
Private or semi-professional use in private office	Commercial use in office or workshop	Industrial use in production or service sector	Industrial use in series production
Application level			
Rapid prototyping "show and tell models"	Rapid prototyping	Rapid manufacturing	Rapid manufacturing "mass production"
Solid imaging Concept modeling	Functional prototypes	Direct manufacturing	Direct manufacturing (series)
Master model for secondary rapid prototyping process	Master model for secondary rapid proto- typing processes	Rapid manufacturing/ direct tooling	Rapid manufacturing/ direct tooling

Computer and CAD skills are of advantage. Concerning infrastructure, there are no specific requirements; see Section 3.2.2 "Professional Printer".

Production printers or *shop floor printers* are machines with the focus on uniformly high-quality parts. This mostly includes a large build space, a (partly) automated material handling system, and more additional equipment for finishing.

Requirements are reproducible processes and parts of high quality. Productivity takes a more important role compared to professional 3D printers. The machines are bigger and release normal emissions in relation to non-additive production machines. Their implementation therefore requires a workshop and a supported extended installation effort. The operators need intensive training by the supplier.

Industrial printers are flexible manufacturing systems, called *flexible AM systems* (FAMS). They provide a uniformly high part quality, similar to *production 3D printers*, and a high output quantity.

The productivity is of primary importance. Manual handling has to be eliminated or at least reduced to a minimum. Items for the supervision and control of the process are integrated.

The operation is free of human interaction. Parallel operation of multiple machines operated by one operator is possible. All this requires a production infrastructure.

While, with regards to these four categories, a broad consensus has been reached, the used terms vary according to the companies' strategies. An increasing trend exists to use the term printer.

The above classification is based on the technical abilities of the machines and not directly on the additive manufacturing processes according to which the machines are working, and which are discussed in Chapter 2 "Additive Manufacturing Processes/3D Printing". Nevertheless, a loose connection exists between the additive manufacturing process and the application level.

Numerous types of machines are presented after discussion of the basic principles in Chapter 3 "The Additive Manufacturing Process Chain and Machines for Additive Manufacturing".

1.3.2 Classification of Machines and Properties of Parts

There is a correlation between machine classes and the associated parts which can be manufactured by these machines and the part properties.

Personal 3D printers predominantly produce mock-ups or parts not subject to load with limited geometrical complexity and low accuracy regarding details.

Professional 3D printers primarily generate mock-ups or low-loaded parts, but nevertheless achieve a significantly better accuracy of details compared to personal printers. Due to the accuracy of details, parts manufactured by professional printers are often converted to high-quality final parts by subsequent processes.

Production 3D printers are used for additive manufacturing of one-of-a-kind parts or (small) series of different parts and to produce final products after external process-determined finishing.

Industrial 3D printers are used for additive manufacturing of quality parts. These printers integrate (at least partly) the process-related post-processing and produce final parts. The output ranges from "one-offs" and a low mix of identical parts to a high mix of different parts. Today (2018), this tendency can be observed for both metal processes such as Additive Industries' "MetalFAB 1" and plastic processes like 3D Systems' "Figure 4".

This correlation between the machine classes and the parts which can be manufactured by these machines does not follow any set pattern. There are simple plastic clips produced by fabbers that are used as final parts, and complex parts manufactured by production printers, e.g. turbine vanes with internal cooling channels, which can only serve for demonstration purposes.

■ 1.4 Conclusions

The discussion about the application of additive manufacturing shows that nearly all application levels and all branches benefit from the options of additive manufacturing. The mentioned definitions support technical discussions. In practice, it is particularly important to differentiate between the various application levels. Disappointment of users frequently is based on insufficiently defined requirements.

The examples clearly show that multiple additive manufacturing processes are available to meet the requirements of the users – to some extent, even alternative solutions exist. These alternatives, the various AM processes available on the market, are presented and discussed in Chapter 2 "Additive Manufacturing Processes/3D Printing".

Today's restrictions of additive manufacturing processes such as limited material choice, insufficient surface quality, inconsistent part quality, or low production output will be continuously reduced. A great number of scientists and industrial product developers worldwide are working on the improvement of all issues of this new technology.

First approaches for products from multiple-composite material have already been presented. This will enable new application fields for all kinds of industrial production, especially for electronic parts and medical applications (see Chapter 4 "Applications of Additive Manufacturing").

■ 1.5 Questions

1. **What are the essential characteristics of parts which were generated by means of additive manufacturing?**

   Nearly any geometry, whatever can be generated, is possible. Due to the layer-oriented manufacturing process, all additively manufactured parts show a stair-stepping pattern on their surfaces.

2. **Why do technology levels and application levels have to be distinguished?**

   Technology levels describe the theoretical and scientific background, while application levels define the use and the AM application, as well as the expected benefit.

3. **What is the difference between generative manufacturing and additive manufacturing?**

None, as these terms are only different expressions for the same technique. By adding and connecting of defined volumes (voxels, layers), parts are generated.

4. **What is the correlation between additive manufacturing and layer-oriented manufacturing?**

Additive manufacturing is a generic term that indicates that parts are built by adding of material or defined volumes. The technical realization of additive manufacturing is solely based on the principle of layer-built elements and is therefore called layer-based technology, layer-oriented technology, or even layer technology. All three terms characterize identical processes.

5. **What are the application fields of solid images?**

Solid images are used to evaluate the general appearance, shape, and haptic properties of parts during project development. They may be used also for the verification of product data.

6. **What is the difference between solid images and functional prototypes?**

Solid images are three-dimensional visual representations, while functional prototypes represent one or a limited number of functional properties of the later product.

7. **Why are follow-up or secondary rapid prototyping processes not part of the family of AM processes?**

Because the parts are not generated layer-wise, but by casting by means of a tool (mold).

8. **Why are indirect processes often called indirect rapid prototyping or rapid manufacturing processes?**

Because the attribute rapid is fashionable and therefore the user expects an improvement of economy.

9. **Why has the term rapid tooling not got its own application level?**

Manufacture of tools and tool elements is technically equivalent to the manufacture of metallic final parts. The machines, the process, and the materials are identical. Therefore, both applications belong to the application level rapid manufacturing. Only the CAD design is different, depending on whether a part (a positive) or the mold for the production of the part (a negative) has to be generated. Consequently, there is no indication for its own application level, but rather for a family of similar applications.

10. **What is the difference between manufacturing of functional prototypes and direct manufacturing?**

In manufacturing of functional prototypes (functional prototyping) prototypes are generated, which only show a few selected functional properties of the final product, while in direct manufacturing parts are generated whose properties are completely identical with those of the final product.

11. **Why do manufacturing processes for prototype tools (prototype tooling) range between the application levels manufacturing and prototyping?**

Because the tool is manufactured applying prototype elements and prototype methods, and therefore is considered as a prototype tool. However, parts produced by prototype tools show – at least under special circumstances – series quality.

References

[1] Burns, M.: *Automated Fabrication*, Prentice Hall, Englewood Cliffs, NJ, 1993

[2] Gebhardt, A.; Hötter, J.-St.: *Additive Manufacturing: 3D Printing for Prototyping and Manufacturing*, Carl Hanser Verlag, Munich, 2016

[3] Gebhardt, A.: *Generative Fertigungsverfahren: Additive Manufacturing und 3D-Drucken für Prototyping – Tooling – Produktion*, 4th edition, Carl Hanser Verlag, Munich, 2013

2 Additive Manufacturing Processes/3D Printing

The technical execution of additive manufacturing processes is carried out by means of layer build processes (also known as *direct layer manufacturing processes*). There are five families of layer build processes that are available on the market as additive manufacturing processes. They are based on different methods to generate a solid layer, to form a part by attaching and connecting adjacent layers. The principle was described in Chapter 1 "Basics of 3D Printing Technology". All five process families, including some derivatives, will be described in detail linked to the available machines on the market, and presented with their typical construction elements.

If prototypes and parts are not produced by means of additive manufacturing, but are based on additive processes, they are called *secondary rapid prototyping processes* or *indirect RP processes*. The principle and the most frequently applied variations are briefly explained in Section 1.2.2 "Indirect Processes".

The most important variation, *room temperature vulcanization*, *RTV*, is described in detail in Section 2.2 "Indirect Processes/Follow-Up Processes".

■ 2.1 Direct Additive Processes

The principle of additive manufacturing is very simple and shown in Figure 1.2. It is merely based on virtual 3D CAD data files (solids). The data files are sliced according to a pre-set layer thickness and thus the free-formed surface (in the z-direction) is realized by a number of contoured layers with uniform thickness (see Figure 2.1).

The (physical) manufacturing process (the AM process) is characterized by repeated generation of single layers and connecting them with the previous one.

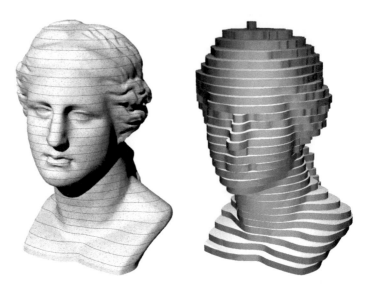

Figure 2.1 Slicing of a 3D freeform area into contoured layers of equal thickness with the
characteristic stair-step effect (thickness exaggerated for clarification)
(Source: GoetheLab, University of Applied Sciences, Aachen)

The process thus consists of two steps, which will be repeated until the part is
finalized:

1. Generation of a single layer, formed in accordance with the contour and thickness defined by the data set for the respective layer.

2. Connecting of each new layer with the previously generated one.

As can be seen from the theoretical representation (Figure 2.1), the growing parts
show the *stair-step effect* – due to the even layer thickness – which is typical for AM
processes (Figure 2.2).

The standard layer thickness is about 0.1 mm, but can be reduced to 0.016 mm
according to the process. Thus, it not only increases the exactness of the parts, but
also the number of required layers, the data volume, and the manufacturing time.

The layer thickness moreover depends on the used material, as the formed "stair-
steps" require more finishing work for hard materials, such as metals and ceramic
materials, as opposed to soft materials like plastics.

Several hundred different types of machines (in addition to countless variations of
fabbers, *DIY*, or *personal printers* (see Section 1.3 "Classification of Machines for
Additive Manufacturing") are on the market. All of them are based on the afore-
mentioned two basic manufacturing steps. They only differ in the way each layer is
generated, how consecutive layers are connected, and the material used.

Figure 2.2 Stair-step effect, layer thickness 0.1 mm; stereolithography
(Source: Gebhardt)

To generate a physical layer, multiple materials like plastics, metals, or ceramic materials in the form of powders, liquids, solids, foils, or sheets can be used. For the process different physical principles such as photo-polymerization, selective fusing, melting, or sintering, cutting, particle bonding, or extrusion can be applied (details in [1] and in Chapter 6 "Materials and Design").

For the outlining of each layer an energy source is required, providing the necessary physical process, as well as a positioning system to follow the x-y coordinates.

The following systems are used:

- Laser with scanner units mostly of the Galvo-type (swiveling mirror devices, optical switches, or gantry-type positioning units (plotters).
- Single or multiple nozzles: print-heads with high-energy lamps, infrared heating devices, or DLP projectors (Digital Light Processing technology).
- Knives or blades in cutting plotters.
- Electron beams with deflection devices.
- Extruder with x-y-z positioning system.

All possible processes can be assigned to five basic families of additive manufacturing processes (Table 2.1). The terms listed in Table 2.1 are so-called generic terms, characterizing all physical principles related to a particular family.

Generic terms have to be distinguished from brand names, given by specific manufacturers to their processes or machines. Both types of terms, inclusive of the common abbreviations, are listed in Table 2.2.

Table 2.1 Physical Principle of Generating and Contouring of Layers and the Resulting Five Basic Families of Generative Manufacturing Processes (Generic Terms)

Generating a solid layer by	Contouring the layer by	Subsequent AM Process
Polymerization	Laser, printhead	Stereolithography Polymer printing
Selective melting or selective sintering and resolidification	Laser, IR-source electron beam	Laser sintering Laser melting
Contour cutting and bonding	Laser, knife, milling	Layer-laminate manufacturing
Selective bonding or gluing by binder	Multi-nozzle printhead	3D printing
Selective application of thermally activated phases	Single-nozzle extruder	Fused layer manufacturing

Table 2.2 AM Processes: Generic Terms, Market Names, and Abbreviations

Generic Name	Abbr.	Brand Name	Abbr.
Polymerization Stereolithography		Laser-Stereolithography Polymer Jetting	SL
Laser sintering Laser melting	LS	Selective Laser Sintering Selective Laser Melting Electron Beam Melting	SLS SLM EBM
Layer laminate manufacturing	LLM	Laminated Object Manufacturing	LOM
3D printing	3DP	Three Dimensional Printing	3DP
Fused layer manufacturing	FLM	Fused Deposition Modeling	FDM

2.1.1 Polymerization

The selective polymerization of a liquid resin (of the epoxy, acrylate, or vinyl ether type) by means of ultraviolet radiation is called (photo-) polymerization. Various processes exist, which only differ in the way the UV radiation is generated, and in the method of contouring. Some polymerization processes result in a partial solidification only. Therefore, a *green part* is produced, requiring additional treatment to achieve a completely cured part. The additional treatment follows after the build process in a special chamber, called a *post-curing oven*. This chamber is equipped with UV lamps, which provide complete and uniform curing of the part.

During the generation of the part, the polymerization processes require supports. These are necessary to stabilize and fix the part, including overhangs, to keep elements in position that are not connected to each other and to avoid deformations like twisting and warping. The supports are added to the 3D CAD model by means of automated software and have to be removed manually after finalizing the complete part. Some types of supports (and materials) can be washed off automatically by special cleaning devices.

2.1.1.1 Laser-Stereolithography (LS)

Stereolithography is the oldest and until today the most detailed AM process. Stereolithography was invented by 3D Systems, Rock Hill, SC, USA. Respective machines were developed and commercialized. By means of laser stereolithography, parts with excellent surface quality and fine details are manufactured. The parts are generated by polymerization of the liquid monomer. With the help of a UV-laser beam the liquid will be transferred into a solid state (solidified) by polymerization. Thus, corresponding to the part's contour from CAD, scaled solid layers are formed. The laser beam is guided by a Galvo-type scanning device, which is directed according to the contour of each layer. A typical machine is shown in Figure 2.3.

Figure 2.3 Laser stereolithography machine from 3D Systems
(Source: 3D Systems)

A laser stereolithography machine consists of a build space (build volume), filled with liquid resin (build material), and a laser scanner unit, arranged on top, which generates the contour in the x-y direction (build area). The build space contains the movable build platform, which can be lowered in the build direction (z-direction; Figure 2.4).

The part is built on the platform. The laser beam generates the contour simultaneously providing the curing of each layer and its connection to the previous one. The movement of the laser beam is directed by a laser scanner according to the contour data of each layer. As soon as the laser beam penetrates the surface of the resin, an instantaneous solidification by polymerization is realized.

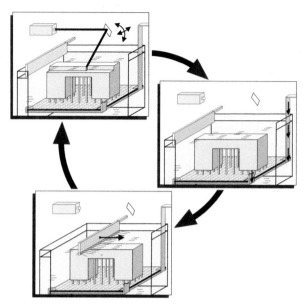

Figure 2.4 Polymerization: laser stereolithography, solidification of a single layer, lowering of the build platform, recoating (clockwise, starting from top left) (Source: Gebhardt)

Depending on the reactivity and transparency of the resin, the layer thickness can be adjusted by the laser power and the tracking speed of the laser beam.

After solidification of the layer, the build platform, including the partially generated part, will be lowered by the amount of the layer thickness. Afterwards the next layer of resin is applied. This procedure is called recoating. Due to the viscosity of the resin, recoating is supported by a leveling system, e.g. *viper* of 3D Systems (if necessary supported by vacuum). The new layer will then be solidified according to its contour. The procedure is continued from bottom to top until the part is completed.

The build process requires supports (Figure 2.5, right), limiting the options for the orientation of the part in the build chamber, as the supports will leave marks on the surface of the part after their removal. Therefore, the orientation should be chosen carefully. Due to the supports the parts cannot be stacked into each other, which reduces the packing density in the build space and thus the productivity.

After the part is finished it is cleaned and finally fully cured in an UV chamber (post curing oven). This step of the process is an integral part of the AM process, called *post-processing*.

The fully cured parts can be shot blasted, polished, or varnished, if required. These process steps are called finishing, which is a measure independent from the AM process. Type and scope only depend on the requirements of the user with regard to the parts and possible restrictions regarding their application.

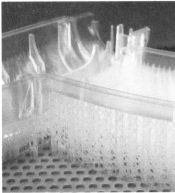

Figure 2.5 Laser stereolithography: thin-walled, shell-type parts (left); parts with supports on the build platform (right)
(Source: CP-GmbH)

Available materials are filled and unfilled epoxy and acrylic resins. Unfilled materials show a comparably low strength and resistance against temperature-related deformations. This low performance can be improved by adding of microspheres or rice-grain-shaped geometric grains made of glass, carbon, or aluminum. This filler material today also contains nanoparticles from carbon or ceramic material.

Typical parts are concept models as shown in Figure 4.10 or thin-walled, shell-type geometries such as a housing for a drilling machine or a hair dryer (Figure 2.5, left).

2.1.1.2 Polymer Printing and Polymer Jetting

When monomer material is applied by print heads, the process is called polymer printing or polymer jetting. The process, which was invented and developed by Objet, Rehovot, Israel, has been commercialized by Stratasys to date.

It could be considered as a 3D printing process, but due to the generation of parts by UV-solidifying of liquid monomers, it represents a polymerization or stereolithography process.

The design of the machines resembles a 2D office printer (Figure 2.6, left). The build material is applied directly to the build platform by means of a multiple-nozzle piezo-electric print head. Simultaneously the solidification is carried out by a twin light curtain, which is produced by two high-performance UV lamps, moving synchronously with the print head. The layer thickness is 0.016 mm (standard) only, providing a very smooth surface. Further layers are generated by moving the build platform in the z-direction. The process is continued layer by layer.

The parts have to be supported while being printed. The supports are automatically generated and simultaneously manufactured by a second set of nozzles by printing. Thus, every layer contains build as well as supporting material.

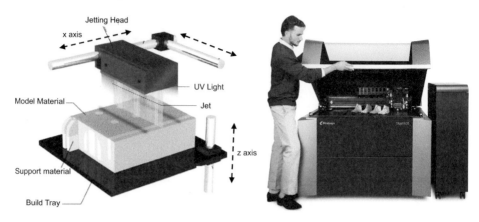

Figure 2.6 Polymer printing or jetting (PolyJet Stratasys): schematic design (left); machine for multiple materials, Objet 500 Connex 3 (right)
(Source: Stratasys)

Consequently, a significant portion of material is consumed for the rigid supports. The supporting material can be washed off in a largely automated finishing process, without leaving marks on the part. A typical part can be seen in Figure 2.7 (left).

Figure 2.7 Polymer printing or jetting (PolyJet Stratasys): helical planetary gears, functional integration (left); wheel of two materials with compressible tire (right)
(Source: GoetheLab, University of Applied Sciences, Aachen/Stratasys)

By means of photo-sensitive monomers plastic parts are manufactured in the process. This facilitates materials in various colors and different shore-hardnesses.

By means of a proprietary technology called PolyJet matrix – combined with a family of machines named "Connex" – parts can be manufactured consisting of two different materials, which are similar to two-component injection molded parts (Figure 2.7, right). This opens up the possibility to manufacture parts from combinations of different materials. Typical parts are thin-walled, detailed, and including internal cavities. Besides the combination of different materials finishing of printed parts is possible. For example, the surface of printed parts can be galvanized – see Figure 2.8 – or adapted to the requirements.

Figure 2.8 Polymer printing: printed car-model, additional galvanizing
(Source: GoetheLab, University of Applied Sciences, Aachen)

2.1.1.3 Digital Light Processing

This process first was presented and commercialized under the name *Perfactory* by EnvisionTEC, Gladbeck, Germany, where the international headquarters is still situated, while the US headquarters is in Detroit. Today the company offers a family of specialized 3D printers amongst which is a bioplotter. Most of them originate from the basic process first commercialized as Perfactory.

This variation of photo-polymerization works with a commercial DLP (*digital light processing*) projector as UV-light source. The complete contour of the current layer is projected onto the part and simultaneously the solidification is initiated.

The projector is mounted in the lower section of the machine housing. The resin is stored in a container with a glass bottom which is positioned above the projector. The current contour of the next layer is projected from below to the lower side of the glass bottom and thus to the resin reservoir. A build platform with an upside down arrangement (Figure 2.9) is immersed from above into the resin reservoir in such a way that between the build platform or the previous layer and the transparent glass bottom there is clearance for the next layer thickness. After solidification of the current layer the build platform is incremented by the amount of the layer thickness to provide the space for the material of the next layer. Due to decreasing pressure in the gap, the fresh material is forced to flow into the void. Due to the small material reservoir, the process is suitable for small parts. It also typically allows quick material changes. The manufacturing of parts requires supports made from the same material.

A large variety of photo-sensitive plastics, including biocompatible types, which can be used for the production of hearing aid housings or master models for dental prosthesis, is available.

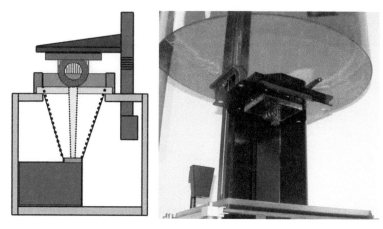

Figure 2.9 Digital light processing, Perfactory, EnvisionTEC: schematic arrangement and resin reservoir with glass-bottom as projection area (left); overhead arranged build platform (inclusive part, right) (Source: Gebhardt/EnvisionTEC)

2.1.1.4 Micro Stereolithography

A large number of processes allows the manufacture of parts in the micrometer range and below (Figure 2.10). Many of them still are in the state of scientific development or on the threshold of industrialization.

Industrially applicable processes use (micro-) laser stereolithography and mask-based systems for mass production of micro parts.

A commercial company that specializes in customized materials and applications and for large series parts, but not currently selling machines, is microTEC, Duisburg, Germany.

Figure 2.10 Micro stereolithography, parts and MEMS (microelectromechanical systems) (Source: microTEC)

2.1.2 Sintering and Melting

Selective melting of plastic and metal powders that show thermoplastic behavior and re-solidification after cooling is called *laser sintering*. Depending on the manufacturer it also is named *selective laser sintering, laser cusing*, or in case of metal *laser melting*. Processes, using an electron beam instead of a laser are called *electron beam melting* (EBM). If the energy is supplied by means of a heat radiator (infrared) and contoured by a mask, the process is called *selective mask sintering*.

2.1.2.1 Laser Sintering/Selective Laser Sintering (LS/SLS)

The terms *laser sintering* and *selective laser sintering* are primarily used for machines that process plastics. Manufacturers include 3D Systems, Rock Hill, SC, USA; EOS GmbH, Munich, Germany; and Prodways, Les Mureaux, France and Plymouth, MN.

The machines of all manufacturers processing metals are very similar. They consist of a build space, filled with powder of a grain size of approximately 20–50 μm, and a laser scanner on top that generates the x-y-contour. The bottom of the build space is designed as a movable piston, which can be adjusted to any z-elevation (Figure 2.11)[1]. The surface of the powder bed forms the build area, where the current layer is generated. The complete build space is heated, to minimize the laser power and to reduce deformations. To avoid oxidation, the build space is flooded with shielding gas.

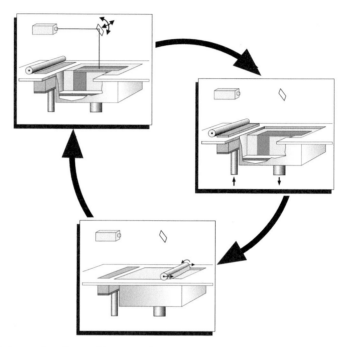

Figure 2.11 Laser sintering and laser melting, schematic procedure; melting and solidifying of a single layer; lowering of the build platform; recoating (clockwise, starting from top left)

Each layer is outlined by the laser beam. The contour data is generated from the slice data of each layer according to the 3D CAD model and managed by the scanner. Where the laser beam hits the powder surface, the particles are locally melted. The process depends on the diameter of the laser beam and on the scan speed. While the beam continues to move the molten material solidifies due to the heat transfer by thermal conductivity into the surrounding powder. Thus, a solid layer is generated.

After the solidification of a layer the piston at the bottom of the build space is lowered by the amount of one layer thickness, which also means that the complete powder bed and the growing part is lowered. The free space above the powder bed is refilled with fresh powder, which is taken from the adjacent powder reservoir by means of a roller.

---

[1] A 3D animation can be found under:
www.rtejournal.de/filme/SLS-RTe.wmv/view

To achieve an even distribution of fresh powder, the roller rotates opposite to the direction of the linear movement of the coating device. This procedure is called *recoating*.

After the recoating the build process starts again, generating the next layer. The complete process is continued layer by layer until the part is finalized. Usually the top layer is manufactured by using a different scan strategy to increase the strength.

After the manufacture is completed and the top layer has been generated, the complete part inclusive the surrounding powder is covered with some additional powder layers. This so-called powder cake has to be cooled down before the part can be taken out of the surrounding powder and be removed. The cooling proce-dure can be carried out inside of the machine; however, cooling down in a separate chamber allows the immediate start of a new build process.

Sintering allows the processing of all kinds of materials like plastics, metals, and ceramics. Basically, the machines are very similar in design. The machines are either adapted to the different materials by software modifications (and possibly by minor hardware modifications) or special versions of a basic machine type are optimized to process a distinctive kind of material.

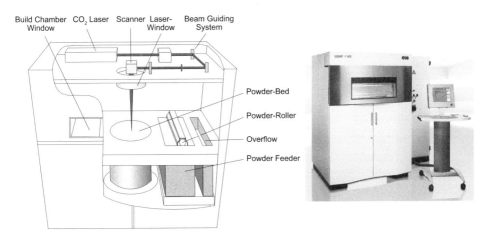

Figure 2.12 Basic design of a sintering machine from 3D Systems (left), laser sintering machine EOSINT P 800 (right) (Source: 3D Systems/EOS GmbH).

In this context the recoating systems are specially adapted to handle the materials, chosen for the application, for example roller systems for plastic powders as well as hopper systems or filling shoes for plastic coated foundry sand. For metal pro-cessing systems, also wiper-type devices are applied.

While standard plastic materials are polyamides of type PA 11 or PA 12, today high-performance materials emulate the properties of PC, ABS, and PA (6.6) plastics, and generate design elements like film hinges and snap-fits. The high-temperature system EOS/P 396 (2017) is the only available system on the market that also processes high-performance plastics (in this case PEEK) and is a trend setter.

For *laser sintering*, unfilled materials and materials filled with spherical or egg-shaped glass, aluminum, or carbon particles, which increase their strength and temperature resistance, are on the market. Flame-resistant materials are also available.

The removal of the part from the powder (the so-called *break out*) is done manually by using vacuum cleaners, brushing, or shot blasting on low pressure.

Semi-automatic break-out stations make work easier and mark the trend to automated cleaning. Metal parts have to be cut manually from the build platform and the supports, which is time-consuming and requires manual skills.

Plastic parts often are porous and need to be infiltrated. If required, surface treatment can be applied, or they can be varnished. Typically, metal parts are dense. Depending on the type of material, they can be processed conventionally, e.g. by cutting or welding.

Sintered plastic parts have properties that are close to those of plastics injection molded parts. They are manufactured either as prototypes (Figure 2.13, left) or as (directly manufactured) parts (also called target parts or series parts) (Figure 2.13, right).

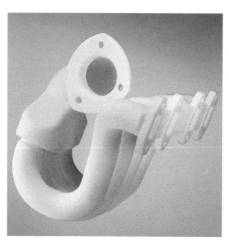

Figure 2.13 Selective laser sintering, SLS (3D Systems): engine exhaust manifold; prototype, polyamide (left); fan rotor, final part (right)
(Source: CP-GmbH)

2.1.2.2 Powderbed Fusion (PBF), Selective Laser Melting (SLM)

Basically, *laser melting* is very similar to the above-described laser sintering process. It was specially developed for manufacturing of very dense (> 99%) metal parts. The material is totally melted by the laser. Therefore, it generates a local (selective) melt pool, which results in a completely dense part after solidification. Generally, the process is named *selective laser melting* (SLM).

Some proprietary terms exist, such as *cusing*, a combination of the words *cladding* and *fusing*.

Most of these machines have their origin in Germany:

- EOS GmbH, Munich
- Realizer GmbH, Borchen
- Concept Laser GmbH, Lichtenfels
- SLM Solutions, Lübeck
- Some of these companies have been acquired by US-American companies like GE (General Electric)

    Moreover, 3D Systems, Rock Hill, SC, USA offers the ProX series, re-branded systems based on the PHENIX process, which is called direct *metal sintering (DMS)*.

    Renishaw developed and offers the machines AM 125 to AM 500, which were taken over from MTT, Great Britain, under their own name.

    For all metal processing machines, a wide variety of metals, including carbon steel, stainless steel, CoCr-alloys, titanium, aluminum, gold, and proprietary alloys is available. Typically, metal parts are final parts and are used as (direct manufactured) products or elements of such products. Characteristic examples are the internally cooled *cooling pin inserts* from tool steel shown in Figure 2.14 (left), which are used for local cooling of injection molds, and the micro-cooler made from AlSi10Mg in Figure 2.14 (right).

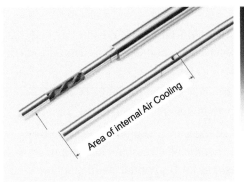

Figure 2.14 Selective laser melting, SLM: internally cooled pin for application in injection molds (left); micro-cooler, made from AlSi10Mg (right)
(Sources: Concept Laser GmbH (left); EOS GmbH (right))

The design of the machines resembles to a large extent machines for laser sintering of plastics. To process reactive materials like titanium or magnesium, fiber lasers with extremely high beam quality as well as sealed build spaces, which are evacuated or filled with shielding gas, are used. Integrated additional heating elements, heating build platforms, are integrated to avoid warping and distortion of the parts.

Machines for micro laser sintering of metal or ceramic parts are still in the development phase. The market introduction of such machines, based on developments of 3D Mikromac, Chemnitz, Germany, was announced by EOS. The typical layer thickness ranges from 1 up to 5 µm; the lowest wall thickness is > 30 µm. A fiber laser with a focus diameter < 20 µm is used. As examples demonstration parts are shown in Figure 2.15.

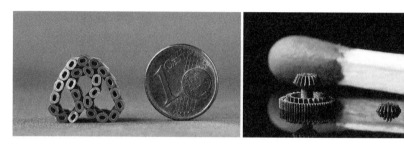

Figure 2.15 Micro laser sintering (EOS): demonstration parts
(Source: EOS GmbH)

2.1.2.3 Electron Beam Melting

For the local melting of material instead of a laser an electron beam can be used. This procedure is called electron beam melting (EBM). As material processing by means of electron beam requires a vacuum, total sealing of the build chamber is necessary.

ARCAM AB from Mölndal, Sweden (recently acquired by GE, Boston, MA) offers a family of EBM machines designed for special applications, e.g. in aviation, medicine, or tool manufacturing (Figure 2.16).

The electron beam provides a high penetration depth and the concept allows a very high scan velocity, which simultaneously can be used for pre-heating.

Therefore, the process works very quickly at increased temperatures. According to statements of the manufacturer, this leads to reduced stress and distortions as well as very good material properties. As example an individual skull implant made from titanium by means of EBM is shown in Figure 4.29 (right).

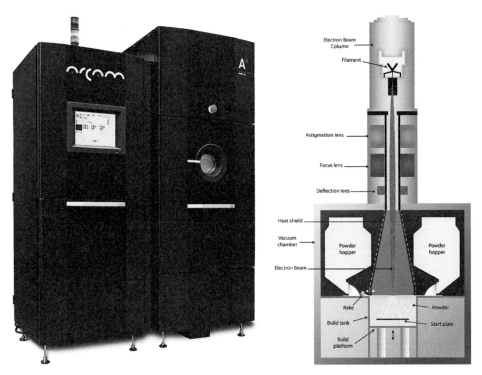

Figure 2.16 Electron beam melting (EBM): ARCAM A2X (left); EBM functional scheme (right) (Source: ARCAM/GE)

2.1.3 Extrusion/Fused Layer Modeling

The layer-wise application of extruded pasty plastics is called *fused layer modeling* (FLM). The process works with pre-fabricated thermoplastic material, which is melted in heated nozzles and applied in strips.

By use of colored material, colored parts can be generated. From technical point of view FLM is an extrusion process (see Figure 2.17)[2]. The parts have to be supported during manufacture.

---

[2] A 3D animation can be found under:
http://www.rtejournal.de/filme/FDM-RTe.wmv/view

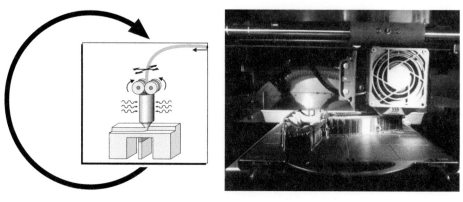

Figure 2.17 Fused layer modeling: extrusion process, functional scheme (left); build space
with build platform, growing part, print head (right)
(Source: GoetheLab, University of Applied Sciences, Aachen)

Many of the so-called *fabbers* or *personal printers* – see Section 1.3 "Classification
of Machines for Additive Manufacturing" – are working with a simplified extrusion
process, most of them without the possibility to provide supports. The most popular
today are Makerbot Replicator+, Ultimaker 3, and Zortrax M200. This branch of the
industry is growing rapidly and is supported by many quasi-private suppliers and
their internet blogs.

Fused Deposition Modeling (FDM)

FDM is a registered, protected brand name for a *fused layer modeling* process, offered by Stratasys Company, Eden Prairie, MN, USA. As it was the first FLM process
on the market worldwide, the term FDM is often used equally to FLM – even as a
generic term.

A FDM-machine consists of a sealed and heated build space (approximately 80 °C
for the processing of ABS plastics), equipped with an extrusion head and a build
platform. The machine works without a laser. The extrusion head provides, dependent on the contour of the current layer, the corresponding material deposition in
the x-y area. It is a device similar to a plotter.

The build material, a prefabricated, thin plastic cord (filament) is continuously fed
into the extrusion head. The material is partly melted by electric heating and
extruded through a nozzle, which defines the filament diameter. Usually the
diameter ranges between 0.1 and 0.3 mm. The build platform is moved in the
z-direction to define the layer thickness, while the material is applied on the top
area of the partly finalized part. The build process needs supports. These are generated by a second nozzle, which simultaneously with the build material extrudes
a different plastic material.

After the application, the pasty material (build material and supporting material) solidifies due to heat conduction into the previously applied layer of the part and forms another solid layer. Then the build platform is lowered by the amount of the layer thickness, before the next layer is applied. This procedure is repeated, until the part is finalized.

The prefabricated wire-like build material is supplied rolled-up in a cartridge. The cartridge is equipped with an integrated sensor that communicates with the material feeding system of the machine and provides the basic information of the material type and for the calculation of the remaining processing time with the corresponding cartridge.

There is a large variety of machines emulating the FDM process. The range covers *personal printers* like the Replicator Mini+ (approximately €1000, 2017), the µPrint (approximately €10,000), the nearly twice-as-expensive *office printer* Dimension, and ranges up to high technical systems like Fortus Production Systems, including the Fortus 900mc, offering the currently biggest build space available on the market (914 × 610 × 914 mm).

For the FDM process numerous plastic materials are available, including construction materials like ABS, PC-ABS, PC, and special developments for medical reproductions. Some machines are limited to a reduced range of different materials. A large number of colors is available, even including fluorescent or translucent, black or white patterns.

As the color is linked to the material, it cannot be changed during the build process (Figure 2.18, left).

Fortus 400 and 900 machines process the high-temperature thermoplastic polyphenylsulfone (PPSF/PPSU). These were the first machines on the market that were able to process this high-performance plastic.

Characteristic properties of FDM parts resemble those of injection molded parts. They exhibit anisotropic behavior, which can be reduced by adjustment of suitable build parameters and proper preheating.

The manufactured parts are used as concept models, functional prototypes, or as (directly manufactured) products.

Due to the extrusion process, FDM parts show typical surface structures (Figure 2.18, center). Depending on the layer thickness and the alignment of the part in the build chamber, these structures are more or less visible. Therefore, the arrangement (alignment) of the parts in the build chamber has a significant influence on their properties and appearance.

Figure 2.18 Fused Deposition Modeling: planetary gear assembled from multi-color FDM parts made from ABS plastics (left); part with supports after manufacturing (center); the same part after removal of the supports and polishing by hand (right) (Sources: Stratasys (left); GoetheLab, University of Applied Sciences, Aachen (center and right))

The post-processing means removal of supports, which can be carried out manually or by means of a special washing device with caustic soda. Polishing requires manual working skills and needs time, and results in high surface quality and surprisingly good results (Figure 2.18, right). Of course, intensive polishing reduces the accuracy of the parts.

In Figure 2.19 (left) earpieces for spectacles are arranged virtually in the build space of a machine.

The mechanism for fixing the earpieces on the spectacles frame is delicate, and therefore the alignment in the build space is important. The side pieces have to be aligned in a way that in delicate areas a minimum of support material is required, and thus less post-processing is necessary. Figure 2.19 (right) shows the printed earpieces, mounted on the frame.

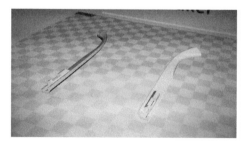 

Figure 2.19 Fused Deposition Modeling, FDM: spectacles earpieces – virtually arranged in the build space of the machine (left); printed and mounted on the spectacles (right) (Source: GoetheLab, University of Applied Sciences, Aachen)

2.1.4 Powder-Binder Process

The layer-wise bonding of granules with a grain size in the 50 μm range by selective injection of liquid binders into the surface of the powder bed is called *three dimensional printing, 3DP*. The process family also is named *drop on powder process*. The process was developed and protected in the early 1990s by MIT[3]. Licenses were granted to Z-Corporation (today 3D Systems) and others, which commercialized the process. Today, variations processing plastics, metals, or ceramic materials are on the market. Most of them work in a two-step process, because after the build phase (first step) the part has to be infiltrated (second step). In the first step, some processes, especially metal processing systems, produce a *green part*.

From this *green part* the binder will be removed by thermal treatment and by subsequent sintering, the final mechanical properties are achieved. As suitable binders are to be found for each kind of powder, the range of materials is nearly unlimited, including applications in the fields of food and pharmaceutical products – though only a small segment currently is commercialized.

Three dimensional printing or more frequent *3D printing* is becoming the synonymous term for all AM processes. Meanwhile it is used as generic term, as 3D printing reminds of two-dimensional printing and thus in a simple way the meaning of AM is expressed. However, the use of two terms with equal status but different meaning sometimes causes confusion. Therefore, especially beginners should avoid mixing the two meanings of the term *3D printing*.

2.1.4.1 3D Printer – 3D Systems/Z Corporation

The machines, developed by Z-Corporation, Burlington, MA, USA (today: 3D Systems) work exactly according to the original 3D printing principle. The process produces *green parts*, which are correct regarding dimensions, but have to be infiltrated to get their strength. The part remains in the powder bed until the build process is finalized. The part is stabilized by the surrounding powder and therefore does not need supports. As the binder can be colored, colored parts can be manufactured accordingly. The company offers a family of 3D printers. Most of them are suitable for the production of continuously colored parts.

The design of the machine strongly resembles a machine for laser sintering. The lower part of the machine contains the build chamber and stores the powder, having a movable piston for the adjustment of the layers and a roller for recoating. On top, a plotter-type device with a commercial print head is arranged, which is similar to a 2D office printer.

---

[3] Massachusetts Institute of Technology, Cambridge, MA, USA

The print head travels across the build area and injects the binder according to the current contour onto the powder. The layer-forming particles are bonded and become one layer of the part. The loose powder supports the part. In contrast to sintering, this process requires neither heating nor shielding gas.

After solidification of the layer, the piston including the complete powder cake and the incomplete part is lowered by the amount of the layer thickness. The space thus formed above the powder cake is refilled with fresh powder, taken from the adjacent powder storage bin and distributed by blade or roller (recoating).

For manufacture of parts and molds and dies for investment casting, starch-like powder and plaster-ceramic grades are available. As the binder can be colored, completely continuously colored parts can be produced, as with colored prints produced by a 2D printer. The production of colored surfaces with texture is a unique sales feature of this process.

Figure 2.20 3D printing: powder–binder process, colored parts (above left and right); machine, Z-printer 450 (below)
(Sources: GoetheLab, University of Applied Sciences, Aachen (above left and right); 3D Systems (below))

After finalization of the last layer, the build process is completed. As the process works at room temperature and the powder cake is at room temperature, powder and the part can be removed from the machine directly after the print. The loose powder is removed by vacuum cleaner, soft brushing, and blowing with air at low pressure.

In a next step the part needs to be infiltrated to improve its strength. This is achieved by using wax or epoxy resin. The result with respect to stability not only depends on the material, but also on the quality of the infiltration. Frequently bubbles are enclosed. Therefore, parts manufactured by 3D printing are not meaningful in stress tests.

Typical parts are concept models (Figure 2.20, above left and right). Z-Printers are of advantage for bureau application and easy to operate (Figure 2.20, below). The surface quality is in comparison to polymerization processes coarse (Figure 2.21, left), but can be significantly improved by manual post-processing (Figure 2.21, right)[4].

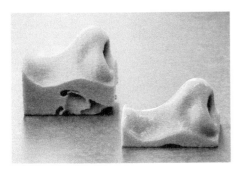 

Figure 2.21 3D printing: powder-binder process; improved surface quality by manual polishing; parts after removal from the machine (each left) and after manual polishing (each right)
(Source: GoetheLab, University of Applied Sciences, Aachen)

2.1.4.2 Metal and Sand Printer – ExOne

ExOne offers 3D printers of the powder-binder type for the manufacture of metal parts and sand cores and molds. Today the focus lies primarily on the marketing presentation of the printing technology and only in the second line on the application.

The metal printing process generates metal as well as metal-ceramic parts by bonding powder by means of micro droplets from a multiple-nozzle print head; see Figure 2.22 (left).

---

[4] A 3D Animation can be found under:
http://www.rtejournal.de/filme/3DP-RTe.wmv/view

The nozzles have a diameter of 46 µm. The continuously emitted binder flow is interrupted by a swinging piezo-element (60 kHz), forming droplets with approximately 80 µm diameter. Before hitting the powder, each droplet passes through an electrostatic field, acquiring an electric charge, which is used for precise positioning.

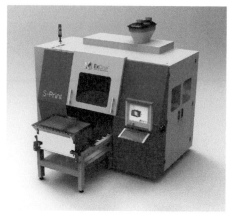

Figure 2.22 Binder jetting: developed at MIT (left); ExOne S-print (right)
(Source: ExOne)

First a model, called a *green part*, is generated in the machine, which gets its "green strength" not by thermal melting, as is the case for sintering, but by injection of a binder into the metal powder. The advantage is that by separation of build material and binder, destabilization of the consistency within the powder does not occur. Moreover, the process works in the machine at room temperature and hence "cold" with minimum distortions.

Figure 2.23 ExOne: foundry sand printer S-Max
(Source: ExOne)

The binder only provides the dimensional stability of the part. By irradiation with a high-energy lamp each layer is dried and solidified, to achieve a transportable *green part.*

The ready-to-use part is generated by de-binding and sintering to 60% of the final density, followed by infiltration (e.g., with epoxy resin) and post-processing like shot-blowing and polishing.

The sand printing machine commercialized by ExOne under the name S-Max belongs to a family of machines that were designed for generating complex cores from foundry sands. The reproducible manufacture of complex cores increases the productivity of sand casting not only for prototypes and test casts, but also for production. The big machine is capable of supplying one production line and therefore has to be considered as foundry machine. S-Max is displayed in Figure 2.23.

2.1.4.3 3D Printing System – Voxeljet

A 3D printing process for plastics and polymer-bound foundry sand – based on the 3D printing principle – was developed by Voxeljet technology GmbH, Friedberg, Germany. The company offers a family of printers, which also includes VX500 (build space 500 × 400 × 300 mm) and VX800 (build space 850 × 450 × 500 mm). The standard layer thickness is 0.150 mm and can be reduced down to 0.080 mm (VX500). The significant element is the high-performance multiple-nozzle print head with 768 nozzles, which is installed in six simultaneously working printing modules. The machines, operating with PMMA and a suitable solvent-type binder, generate plastic parts directly. These are to be used as functional prototypes. Moreover, they can be considered as excellent lost molds for precision casting due to their low ash residues.

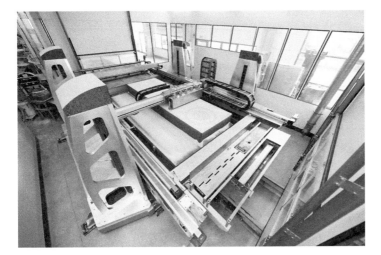

Figure 2.24 3D printing (powder-binder process) : Voxeljet VX4000 plastics printer (Source: Voxeljet)

A concept machine (VX4000) has a very large build space (4 × 2 × 1 m), which allows the manufacture of either a big part or a series of smaller parts in one build course as shown in Figure 2.24. Almost the complete surplus powder can be re-used after finalization of the build.

2.1.5 Layer Laminate Manufacturing (LLM)

Cutting contours out of pre-fabricated foils or sheets of constant thickness according to the 3D CAD dataset and connecting these with the surface of the previous layer is known as *layer laminate manufacturing (LLM)*.

The foils or sheets may consist of paper, plastics, metals, or ceramic materials. As cutting tools, lasers, knives, or mills can be used. Connecting consecutive layers is perfomed by gluing, ultrasound, soldering, or diffusion welding. The majority of the processes only need one manufacturing step; a few require post-processing like sintering in an annealing furnace.

The basic advantage of the LLM process is the fast manufacturing of massive parts.

2.1.5.1 Laminated Object Manufacturing (LOM)

The oldest and worldwide known AM-LLM process is laminated object manufacturing (LOM). It was originally developed by Helysys, USA, currently Cubic Technologies, Torrance, CA, USA. This machine as well as another similar one later developed by Kinergy, Singapore are not manufactured anymore. However, many of these machines are still in use, for which the supplier provides service, maintenance, and contract manufacturing. The build material consists of coiled paper of approximately 0.2 mm wall thickness. The underside of the paper is coated with glue, which is activated during the re-coating process by heating.

The machine contains a build platform, movable in the z-direction, and a device to unwind the paper, position it on the build platform, and roll up the excess paper on the opposite side. A laser cuts out the contours.

To manufacture a part, the paper is positioned and fixed by means of a heated roller, which activates the glue. The contour is cut with a laser plotter, where cutting depth can be adjusted in accordance with the paper thickness. An additional laser-cut designed as a frame defines the borders of the part. Beside the frame, paper strips remain at each side. These strips allow lifting of the unused paper and for it to be wound on the second roller (Figure 2.25)[5]. The material between the contour and the frame remains on the build platform, supporting the part. For easier removal, the waste material is cut into squares during printing.

---

[5] A 3D animation can be found under:
   http://www.rtejournal.de/filme/LLM-RTe.wmv/view

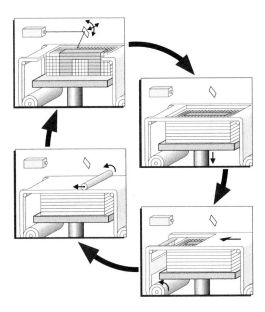

Figure 2.25 Layer laminate manufacturing, laminated object manufacturing (LOM): functional scheme

After the build process is completed, the paper block, including the part and the supporting material, is removed from the build platform. The frame and the squares, which end up as small blocks, are removed, and after varnishing finally the part is ready. Typical parts are gear housings as shown in Figure 2.26.

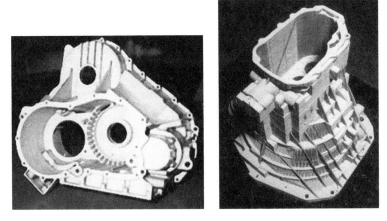

Figure 2.26 Layer laminate manufacturing; paper lamination (LOM): gear housings (Source: VW)

2.1.5.2 Selective Deposition Lamination (SDL)

Mcor Technologies, Ardee, Ireland has commercialized a machine operating according to *selective deposition lamination (SDL)*, which basically is a LLM process. The printer uses standard paper (A4, 80 grams or "letter standard office") or that from a paper reel. Instead of a laser-cut, the contour is outlined with a blade made from tungsten carbide. The process uses commercial office-quality paper and white standard glue of the polyvinylacetate (PVA) type. As this glue tends to corrugate the paper, a special coating system, based on micro droplets, was developed.

The amount of droplets is adjusted to a lower level in areas that do not belong to the part, to make cleaning easier. Using colored paper, colored elements can be generated even in otherwise monochrome machines. To get the colored structure of the part, the paper has to be stacked manually in the right sequence.

Mcor IRIS HD is a further development with an integrated color printer which allows colored printing on the edge strip and the edge itself resulting in a colored 3D part after completion of the build (Figure 2.27).

Mcor ARKe (Figure 2.28, left) is the first full-color desktop printer or fabber. Strategically it is aimed at the market of personal 3D printers and an attractive price is offered for young clients, featuring "individual" housing decoration, intuitive handling, smartphone access, and a refreshing product presentation.

Figure 2.27 Selective deposition lamination (SDL): machine Mcor IRIS HD (left); colored skull, made from paper (right)
(Source: Mcor Technologies)

Figure 2.28 Selective deposition lamination (SDL): machines Mcor ARKe (left); view into the build space and full colored part from paper, placed on the machine (right) (Source: Mcor Technologies)

2.1.5.3 LLM Machines for Metal Parts

Most of the LLM processes for manufacture of metal parts work by means of cutting and joining of metal (e.g., steel) sheets.

The contour is generated either by laser cutting or milling of sheets. The sheets are joined by means of diffusion welding, submerged welding, soldering, or mechanically by bolting. This kind of semi-automatic process, consisting of several steps, does not really represent an AM process, although it is an additive and layer-oriented process.

Laminated Metal Prototyping

The principle of manufacturing parts from metal segments is colloquially called metal-foil LOM. It corresponds to the principle of layer milling processes.

It is primarily applied for the manufacturing of tools. The part is completely designed by CAD and sliced into the required number of contour sections. The nested contours are cut by laser from a steel sheet.

2.1.6 Hybrid Processes

An analysis of additive processes shows that often, besides additive manufacturing, primarily subtractive process steps are applied. The LLM processes, for example, use blades or milling tools to contour the layer. In this respect, many additive processes have to be considered, strictly speaking, as hybrid processes (additive/subtractive). To distinguish between additive and hybrid processes therefore is to a certain degree arbitrary.

Hybrid processes in this section are considered as manufacturing processes that link the potential of at least one additive with that of one non-additive process step.

A typical attribute is that exclusively non-additively working machines – primarily milling machines – are upgraded by additively working additional modules (hardware) to hybrid manufacturing plants.

2.1.6.1 Controlled Metal Buildup (CMB)

The machine of Hermle Maschinenbau GmbH, Ottobrunn, near Munich, Germany operates according to *controlled metal buildup (CMB)*. Powder of different metals and alloys is applied locally onto a part by a carrying gas jet via powder nozzles. It is simultaneously locally melted by means of a laser and solidified by thermal conductivity resulting in a rigid layer. A milling device provides the defined geometric shape and surface quality. In this way material is generated line by line or layer by layer, respectively. Thus, a part is built or repaired. In Figure 2.29 a part is shown after the AM process (above) and after finishing by machining (center and below).

Figure 2.29 Hermle MPA40: part after AM process (above), after completion (center), and as a cut-away to present the internal structure (below) (Source: Hermle)

2.1.6.2 Direct Metal Deposition (DMD)

The LASERTEC 65 hybrid machine of the DMG MORI company combines laser weld overlay and 5-axis milling (see Figure 2.30) and thus enables the complete manufacture of complex metal parts. The additive step works according to the *direct metal deposition (DMD)* process. Powders from different metals and alloys can be processed. The process does not need supports and, due to the milling step, the parts can be manufactured without *stair-steps*.

Figure 2.30 DMG MORI LASERTEC 65: processing sequence with powder nozzle (left); processing sequence with milling head (right)
(Source: DGM MORI)

2.1.6.3 Extruding and Milling – Big Area Additive Manufacturing (BAAM)

The development of the BAAM machine is based on a project of the Oak Ridge National Laboratory (ORNL) research center and Cincinnati Inc. (in 2014), and which was sponsored by the National Department of Energy.

The aim of the development is a machine for the industrial production of large parts, primarily made from high-performance plastics (large-scale polymer additive manufacturing system). As a demonstration object, the Strati of Local Motors was manufactured at the IMTS 2014 (International Manufacturing Technical Show), claiming to be the first printed car worldwide (Figure 2.31 and Figure 2.32).

The BAAM printer works according to the extrusion process. It can manufacture parts with dimensions up to x, y = 4 × 2 m.

An extruder polymerizes commercial plastic pellets. The paste is deposited as comparably thick layers (in the mm range) on the part by means of a nozzle. They are squeezed onto the previous layer, in such a way that the single track usually is twice (up to four times) as wide as high, and shows a flat top. This is achieved by a kind of intensively vibrating "trowel", through which the nozzle is extruding onto the part. Thus, the build material is guided between nozzle and part. The "trowel" is moved parallel to the build area by an eccentric shaft to prevent sticking. This procedure is called *tamping*.

Increased wall thickness is achieved by building thinner walls with a single or double track, filled in by a honeycomb or grid structure. This system resembles the *infill* function of personal 3D printers that helps to reduce the manufacturing time as well as the parts weight. The deposited track solidifies by cooling down after heat transfer into the part.

The process needs supports. However, it is the aim of design measures to avoid these. The parts can be post-processed by milling, which is not performed directly in the process.

Figure 2.31 Big area additive manufacturing (BAAM): view into the build space showing the part on a heated build platform. The displayed gantry (moving in the x-direction) handles the movable extruder-nozzle unit on top
(Source: Cincinnati Inc.)

Figure 2.32 Big area additive manufacturing (BAAM): situation at the optional milling while post-processing

2.1.7 Further Processes

2.1.7.1 Aerosol Printing

A very interesting process with great potential is *aerosol printing*, which was developed by OPTOMEC, NM, USA and commercialized as *maskless mesoscale materials deposition (M3D)*.

A stream of very fine droplets (aerosol), consisting of extremely fine particles with diameters in the nano-range, is directed to the surface of a substrate. The aerosol is deposited following a scheme, designed by CAD. The liquid phase vanishes by evaporation, while the particles remain on the substrate.

The particles can consist of ink, metals, ceramic materials, plastics, or even living cells. Depending on the kind of material, post-processing by means of laser may be required.

Aerosol printing is a promising process for electronic elements, but also for the artificial production of human tissue (tissue engineering).

As the process is currently only applicable as 2½D surface structuring on 2½D objects and not yet on real 3D objects, many people do not consider it to be a true AM process.

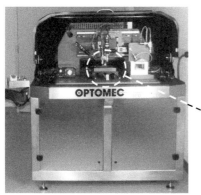

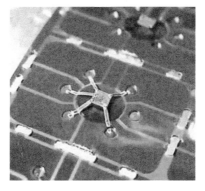

Figure 2.33 Maskless mesoscale materials deposition (M3D): machine and printhead (above); surface structure for a "smart card" (below) (Source: OPTOMEC / eppic-faraday)

2.1.7.2 Bioplotter

As already mentioned in the context of polymer printing machines (Section 2.1.1 "Polymerization" and the OPTOMEC M3D machine, Figure 2.33), one of the special features of these AM processes is the potential to simultaneously process multiple materials. The 3D Bioplotter, a registered brand of Envision TEC, Marl, Germany allows the processing of a large variety of materials, including the plastics polyurethane and silicone, the building material for bones hydroxyapatite (HA), and the medically interesting materials PCL (polycaprolactone[6]), and collagen and fibrin for printing of tissues and organs. Up to five materials may be processed by means of a heated or cooled distribution unit, moved by a plotter, which itself is moving in three axes.

Depending on the material, the system uses different solidification principles as precipitation change state (liquid to solid) and two-component reaction. Some materials need post-processing by sintering.

■ 2.2 Indirect Processes/Follow-Up Processes

As discussed in Section 1.2 "Application Levels", several parts, small series, or parts with improved material quality can be manufactured by counter-casting of a 3D printed precise AM master-model. These kinds of processes are called *follow-up processes or secondary rapid prototyping processes*. The most well-known process is *vacuum casting*, also called *silicone casting* or *room temperature vulcanization, RTV*. The process is illustrated in its fundamental steps in Figure 2.34 and contains in total:

Preparation of the master → definition of the parting line and positioning of the master in the die box → casting with silicone and crosslinking → opening of the silicone mold → removal of the master → preparation and closing of the casting mold for manufacturing of cast parts → positioning of the casting mold in the vacuum chamber and filling in of casting material → casting process with polyurethane → opening of the casting mold and demolding of the part → cleaning of the part and quality inspection.

The same casting mold can be used for multiple casts with different colors or from different materials (Source: MTT Technologies).

---

[6] A biodegradable polyester

Master Model, finished

Definition of Parting Line
Positioning in Die Case

Casting of Molding
Material

Opening of the Mold

Removal of the Master

Casting of the PUR Material
Solidification by Reaction

Removal of the Part
Quality Control

Cast Parts

Figure 2.34 Follow-up process or secondary rapid prototyping process: vacuum casting or
room temperature vulcanization (RTV).
(Source: MTT Technologies)

■ 2.3 Conclusions

Additive manufacturing processes are new outstanding manufacturing technologies with unique features. AM machines and the affiliated processes are capable of processing a great variety of materials. Besides plastics, metals, and ceramic materials, also medical or biological materials can be processed to products. Material can be mixed during the ongoing process and thus different properties within one part can be generated (*graded materials* with locally different properties).

Traditional plastics like polyamide can be processed with nearly no geometrical restrictions. Thus, a lot of design restrictions to which designers today are subject to are eliminated.

Numerous types of machines are on the market and the number of new machines, working with existing or new processes, is increasing steadily.

■ 2.4 Questions

1. **What is the principle of additive manufacturing (AM)?**

   The part to be produced has to be defined by a 3D data set. To exploit the data, the part to be produced is sliced virtually by means of so-called slice algorithms into slices or layers of equal thickness. Thus, the contour data are generated according to the layer thickness.

   The AM machine transfers the virtual data set into real layers and connects each layer with the previous one. Due to the constant layer thickness the part shows a *stair-step* effect on its surface. The entire 3D printing process functions automatically.

2. **Why do additively manufactured parts show a stair-step effect?**

   Because of the constant layer thickness and the uniform contour with regard to the layer thickness.

3. **What are the two essential steps of each additive manufacturing process?**

   Step 1: Generation of a single physical layer with constant layer thickness and an outer shape according to the contour, both based on the slice-data-set.

   Step 2: Connection of each single layer to the preceding one (staggering the layers).

4. **What is the typical layer thickness of additively manufactured plastic parts?**

Most of the machines generate layers in the range of 0.1 mm; however, there are machines producing layers with a thickness of 0.016 mm, as well as machines working with a layer thickness of approximately 0.2 mm and up.

5. **Description of AM processes (selected processes)**

Stereolithography

Polymer Printing or Jetting

Selective Laser Sintering (Selective LS)

Selective Laser Melting (Selective LM)

Laminated Object Manufacturing (LOM)

Fused Deposition Modeling (FDM)

3D Printing

For answers see text.

6. **There are three AM processes to generate colored parts; what are their names?**

Extrusion processes, especially fused deposition modeling (FDM), 3D printing (powder-binder process), and Mcor paper-layer process.

7. **How do the above-named processed differ?**

FDM currently generates just one color at the time, which is provided by the pre-fabricated filament.

With 3D printing (powder-binder process) a continually colored part can be manufactured as is achieved with a 2D office printer.

The Mcor paper process produces fully colored parts by means of a colored 2D printer that marks the edges of the layer thus resulting in a colored surface after completion.

8. **Why should a part manufactured by 3D printing (powder-binder process) not undergo strength tests?**

Because the properties of the part depend more on the quality of the infiltration, rather than on the 3D print process or the material.

9. **How do master models have to be prepared for vacuum casting?**

They have to be polished. Filler pipes for the inlet of molten material and vents have to be provided. The parting line has to be defined.

10. **Which processes can be used to connect metal layers, applying the LLM (layer laminate manufacturing) process for metals?**

Diffusion welding, soldering, submerged arc welding, mechanical joining.

References

[1] Gebhardt, A., Hötter, J.-St.: *Additive Manufacturing: 3D Printing for Prototyping and Manufacturing*, Carl Hanser Verlag, Munich, 2016

3 The Additive Manufacturing Process Chain and Machines for Additive Manufacturing

■ 3.1 Data Processing and Process Chains

A complete and error-free 3D data set is the indispensable prerequisite for additive manufacturing. In most cases the product data set is generated during the design procedure by means of a professional 3D CAD system. Alternatively, a scanner and other processes for the generation of digital images – including also those used for medical applications (e.g. CT, computer tomography/MRI, magnetic resonance imaging) – are applied. Most of these systems have to be operated by skilled personnel, usually by design engineers. Due to the growing use of additive manufacturing by private persons or by professionals whose core business is not in the field of AM, easy-to-handle CAD systems and ready-to-use 3D data sets are of increasing importance.

Although the generation of data sets is a prerequisite for the application of additive manufacturing, it is a separate issue that cannot be discussed in detail in the scope of this book. There are however two aspects that are closely linked to the course of additive manufacturing and therefore are addressed here. First, it is important to exactly know the standard process chain and the data flow between CAD and AM. Second, there should be knowledge of the application level of the final product and the interaction with the AM process chain and the CAD design.

3.1.1 AM Process Chain

As shown in Figure 3.1, the AM process starts with a faultless 3D data set. It contains all (geometric) information regarding the part that will be manufactured additively. Commonly, it is called the *virtual product model*. If the additive manufacturing process requires supports, these are generated by a separate program and further on are treated as elements of the data set.

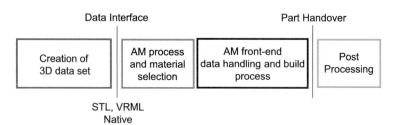

Figure 3.1 AM process chain

Although the generation of the data set is mostly an element of a professional product development process, the data may be obtained by using a 3D CAD program from various sources, such as scanning. As overall requirement, the data set has to represent a closed 3D volume model.

The interface between the generation of the 3D data set and the AM process chain is called the "AM front end". It marks the starting point of the additive manufacturing process.

For the transmission of the 3D data into the additive manufacturing process, the so-called STL format plays the prominent role. Today it represents a de facto "industrial standard" of AM and allows control of nearly all AM machines. The same is valid for the AMF format (Additive Manufacturing File format), which was developed based on the STL format. Because of its basic importance for AM, it is discussed more in detail in Section 3.1.2 "Data Structure, Errors, and Repair". For data transmission, besides proprietary formats of CAD systems (natives), other formats are also used, including VRML (Virtual Reality Modeling Language) and PLY (Polygon File Format), but are significantly less widespread.

After modeling, an appropriate AM material and thus a corresponding AM process is selected, based on the requirements of the part. The material-dependent process parameters for the build process are obtained from the manufacturers – mainly from databases or determined in separate tests. Finally, they are deposited in a material data file. Ambitious users carry out these optimizations on their own, whereas others engage institutes or use manufacturers' data. The communication between operator and machine is supported by icon-based software and simple mouse clicks. This is especially the case for office machines (professional printers).

AM or front-end software, which for historical reasons is often called "rapid proto-typing software", is mostly integrated into the standard machine software that comes with the machine. Alternatively, third-party software (a widely spread program package comes from Materialise) is used. It mostly provides a wider choice of functions and offers useful options especially for professional users.

Before starting the build process, the parts are positioned in the build space and machine-dependent parameters, such as for example the recoating time, are set.

Then the build process is started and continues automatically layer by layer until the part is completed. After completion, the part passes – as far as necessary – through machine-specific cooling procedures and then is taken out of the machine. In some cases, the build process generates a kind of "green part", which has to be treated by further non-additive process steps like infiltration or separate post-sintering. Depending on the process (see Chapter 4 "Applications of Additive Manufacturing") the parts have to be cleaned by solvents or post-processed by separate low-pressure sand blasting. Also depending on the process, supports may have to be removed and additional curing or infiltration might have to be carried out. These process steps are elements of the AM process and are called *post-processing*. Furthermore, the parts can be machined according to the type of material and for example be varnished or flock-coated. These post-processing steps, which are not part of the original AM process, are called *finishing*.

All AM processes basically follow the above-described general AM process chain. Today (2018) the same processes and materials for both application levels *rapid prototyping* – the manufacture of prototypes – as well as *rapid manufacturing* – the manufacture of final products – are applied.

The machines for *rapid prototyping* and *rapid manufacturing* are without exception more or less identical. Whether a part is a prototype or a final product is determined by the data flow and the process chain.

3.1.1.1 Process Chain: Rapid Prototyping

Rapid prototyping parts serve as solid images for the evaluation of the appearance and proportions of an end product, as well as for the evaluation of certain (isolated) properties such as fit (see Section 1.2.1 "Direct Processes"). As explained in Section 3.1.1 "AM Process Chain", the 3D CAD design process is an upstream and independent step before the later AM manufacturing process. The design is carried out in accordance with the design rules of the subsequent non-additively manufactured series products. The AM process and the AM material will be selected after the completion of the design in such a way that the properties of the target part (the later series product) are met as closely as possible, thus effectively supporting the evaluation. The AM machine merely produces 3D objects based on the data set and from printing material and so indeed acts like a printer (see Figure 3.2, above).

The data flow and the workflow reflect the different responsibilities of design and manufacturing (Figure 3.2, below).

The design department takes responsibility for generating the 3D data set. The design is based on the design rules for the later series product and thus the series production process.

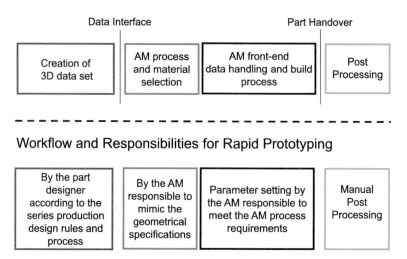

Data Interface

Part Handover

| Creation of 3D data set | AM process and material selection | AM front-end data handling and build process | Post Processing |

Workflow and Responsibilities for Rapid Prototyping

| By the part designer according to the series production design rules and process | By the AM responsible to mimic the geometrical specifications | Parameter setting by the AM responsible to meet the AM process requirements | Manual Post Processing |

Figure 3.2 Data flow and process chain for rapid prototyping

The resulting product data is transferred, according to Figure 3.2 (above), to the AM machine. To produce a prototype on this basis, e.g. for the product evaluation, a suitable AM material and a matching AM process are chosen. Criteria to be observed are that the material should correspond as much as possible to the later production material, and the prototype should represent the essential properties of the later series part (e.g., dimensions, details). For manufacturing by 3D printing, the position of the part on the build platform and its orientation are to be defined, and the build parameters have to be adjusted by means of the software of the machine.

The responsibility for this process step has to be taken by the AM manufacturer. The technological properties of the part are determined by the AM process.

Usually, the finishing – following the build process – is carried out manually.

In summary, rapid prototyping is applied to get a first scaled sample of a later series product made from sample material that mimics the series material in order to evaluate the entire production process. It has to be stated that a prototype manufactured by means of an AM process, although designed in accordance with the design specifications of the later series part and its material properties, is manufactured from a different material by means of a different production process, the AM process. Thus, an AM prototype may simulate a series product, but is not identical with the series production.

The two skull models in Figure 4.28 underline this statement. They are based on identical 3D CAD data sets, but show totally different properties, depending on the choice of the 3D printing process (powder-binder process, left; stereolithography, right), and the AM material (epoxy resin, left; plaster ceramic, right).

3.1.1.2 Process Chain: Rapid Manufacturing

Rapid manufacturing parts are final products according to the AM application level *rapid manufacturing* (see Section 1.2.1 "Direct Processes"). They have to show all material and product properties that have been defined in the design phase, and hence the responsibilities are changing. Design and production cannot work totally independently from each other.

At the beginning of the design work, the designer has to select the build material in accordance with a suitable AM process. In contrast to rapid manufacturing, additive manufacturing is now the final production process. Therefore, the properties and all required parameters do not only have to be defined for the part, but also for the entire build process. This task includes AM-specific activities like scaling, positioning, and aligning of the part on the build platform. Thus, the operator of the AM machine is not responsible for these parameters, but only for the proper operation of the machine.

Regarding rapid manufacturing the responsibility of the designer is extended significantly (see Figure 3.3). It can be stated that the responsibilities of design and manufacturing are overlapping to a large extent and AM needs specialists who have profound knowledge of both fields.

For rapid manufacturing applications, in contrast to (in most cases) rapid prototyping, the productivity is of importance. *Post-processing* and finishing should be automated as far as possible.

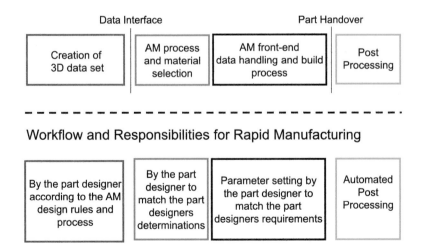

Figure 3.3 Data flow and process chain for rapid manufacturing

3.1.2 Data Structure, Errors, and Repair

The STL format presently still is said to be the (de facto) standard format for data transfer in the field of additive manufacturing.

To generate a STL data set from a 3D CAD file, the inner free-form surface of a volume as well as the outer one is approximated by triangles according to its curvature. Large triangles are used for slightly curved or flat areas and small triangles for strongly curved areas. This procedure is called *triangulation* or *tessellation*.

STL is a very simple data format. Each triangle is defined only by the coordinates of the three-corner points (x, y, z-coordinates) and the normal vector, describing the orientation of the surface. With the help of the triangles, the data set can be sliced easily in any desired z-coordinate, to gain the contour of each layer. STL originally meant *standard transformation language*, and today, due to its close link to additive manufacturing processes, is called *surface tessellation language* (DIN EN ISO/ASTM 52900, VDI 3405) or sometimes *stereolithography language.* The corresponding file extension is .stl. The STL format can be exported directly from almost any 3D CAD system and any scanner software or generated from 3D CAD data by means of rapid prototyping software. STL data sets can be imported and processed by any AM machine. Therefore, STL is the de facto AM standard currently.

STL data sets can be checked by a "view" function that is part of almost any front-end or rapid prototyping software. Moreover, it supports the orientation and the positioning of the part to be manufactured on the build platform as well as the grouping of multiple parts to be built in a mutual procedure (print). It also supports the scaling and calculates the amount and type of the necessary supports. Furthermore, it adds build parameters and, in most cases, carries out the "slicing" of the data set into layers. As a service function, the software mostly estimates the build time and controls the complete build process including material management and, if required, the heating-up and cooling-down cycles.

An advantage of the STL formulation is that the triangles can be varied in size and thus the data set can be adjusted in extent and accuracy, to fulfill the requirements of the clients. A reduction of the number (and thus the enlargement) of the triangles also reduces the extent of the data set, albeit with decreasing accuracy – and vice versa. The principle of generation and the effect of differently sized triangles can be seen in Figure 3.4, considering a geometrical object (sphere) and a technical part (ring).

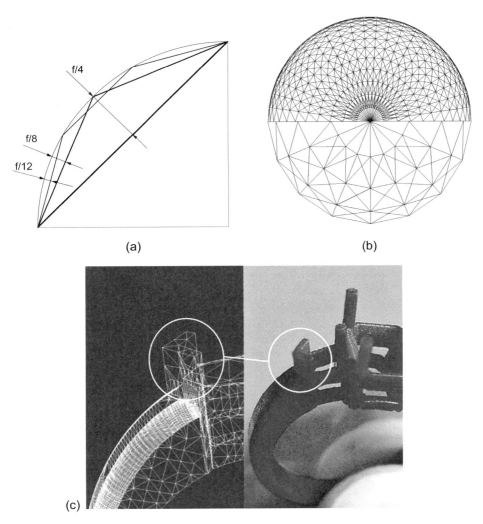

f/4

f/8

f/12

(a)   (b)

(c)

Figure 3.4 Triangulation: (a) principle; (b) effect of different triangle sizes on a sphere; (c) on a technical part in the construction state (left) and on the part (right)

The prominent advantage of the STL formulation is that a triangle can be cut easily at any desired coordinate, providing the possibility to "slice" the STL data set corresponding to the layer thickness, generated by the AM process. Some processes enable adapted "slicing" that also is based on triangulation. "Slicing" based on triangles can be carried out independently of the original 3D CAD data set, which is considered as an additional advantage, as not the complete 3D CAD model, but only the STL model has to be transferred.

Beside STL data sets, different formulations like VRML or contour-oriented direct cut data like SLC can be used in the same way. Other customary file-types are .WRL, .PLY, .MAX, and .SLDPRT. Further details are to be found in the literature, e.g. [1].

STL data sets derived from 3D CAD data sets may include errors that are often a result of careless working with CAD or during scanning. When 3D data sets are frequently exchanged between CAD systems, errors may have their origin here. Data errors are basically not major problems as efficient repair software is available. Usually it is integrated into the software editing the AM data set.

To give the user an idea of the types of errors and their consequences, the most frequent errors appearing in STL data sets are illustrated in Figure 3.5.

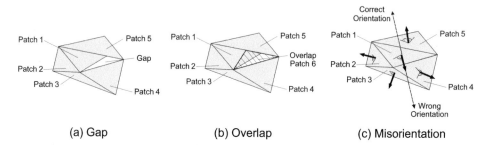

(a) Gap   **(b) Overlap**   **(c) Misorientation**

Figure 3.5 Errors in STL data sets: (a) holes or gaps, (b) overlapping triangles, and (c) wrong normal vector

Due to an unsuitable adaption, adjacent triangles show holes or gaps (a) as well as doubly defined or overlapping triangles, e.g. segments of triangles (b). In both cases (see Figure 3.5) the tool path for the layer cannot be calculated adequately, causing errors. The worst lead to uncompleted parts or the complete termination of the build. Similar problems occur if triangles intersect with each other.

The volume of a body, especially the wall thickness, is defined by the space between two boundary walls. The differentiation between inner and outer wall is made by means of the normal vector. If the normal vector is pointing into the wrong direction, it causes holes, double wall thickness, or some unpredictable geometrical deviations, as shown in (c). All these errors can be detected in an uncomplicated way by demonstration data sets or from simple parts. Real parts are much more complex. Consequently, errors are not evident and difficult to detect. Therefore, the data set should be examined before the manufacture is started.

The "Additive Manufacturing File" (AMF) format is a standard for data sets that besides the geometry also contain information about different materials, colors, textures, and other physical-technological properties.

By means of the AMF format, machines that generate textures, colors, multiple materials, and gradient materials, as well as micro structures directly in the process can also be controlled. It was published by the ASTM (American Society for Testing and Materials) as F2915-12 [ASTM F2915-12].

The AMF format is directed by following main commands in the "header" that are only addressed if the adequate information is available in the data set and can be processed by the machine. These may be applied to the complete part, as well as to single structural elements:

1. <object>

   This command defines one or multiple material volumes. In an AMF data set at least one object must be available; several objects are optional for construction groups.

2. <material>

   Defines materials by IDs and assigns them to destructive volumes.

3. <texture>

   Enables the assignment of color patterns and textures.

4. <color>

   Defines colors by the red, green, blue, and alpha channel. The alpha channel controls the transparency. The <color> element can be applied to materials, objects, volumes, vertices, or triangles.

5. <constellation>

   Defines the sequence of build for parts, if different parts are enclosed in the AMF data set.

6. <metadata>

   Within the <metadata> scope additional information, such as author, project name, copyright, used CAD system, etc., is defined.

The AMF format processes not only straight edges as the STL format does, but also curved ones and thus enables a better approximation of the geometry. This reduces the amount of data, while the resolution of the geometry stays constant.

The AMF format is upwards as well as downwards compatible with classic STL files. The transmitted function however is reduced to the geometry. The AMF format is progressively compatible, allowing future extensions.

The format is XML-based and therefore platform-independent. AMF is open free-ware and thus contributes to preventing a further increase of proprietary solutions. Lastly, one of the biggest advantages of the STL formation is that its application is independent of type of machine.

The AMF format is noticeably increasing in user spread but not yet generally accepted.

■ 3.2 Machines for Additive Manufacturing

As already mentioned in Section 1.3 "Classification of Machines for Additive Manufacturing", the large and fast-growing variety of different printers (machines) can be split into four classes. The four classes or categories of AM machines are *personal printers, professional printers, production printers,* and *industrial printers* (see Figure 3.6).

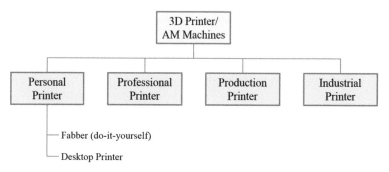

Figure 3.6 Classification of 3D printers/AM machines

In Table 3.1 the four classes of AM machines are assigned to the application levels and classes as well as to the predominantly applied build material.

Apart from the classification of AM machines according to their application levels and fields, they differ explicitly in relation to the used materials, the accuracy (resolution), and the maximal load of the parts, but also with regard to the complexity of the parts.

Table 3.1 Classification of Machines for Additive Manufacturing

	Machine Class			
Designation	Personal printer	Professional printer	Production printer	Industrial printer
Application	Private/semi-professional	Professional	Professional/(industrial)	Professional/industrial
Build material	Plastics	Plastics, metal	Plastics, metal, ceramic	Plastics, metal
Application Level (see Figure 1.1)				
Prototypes	X			
Concept models	X			
Functional parts		X		
Final products			X	X

The different AM classes do not show a fixed relation to the applied AM process. To illustrate, at least by trend, the options and limits of the machine classes, they are discussed in the following sections.

3.2.1 Personal Printers

3D printers manufacture complete parts (products or elements of products) automatically and in one step. This enables also non-skilled (production) workers to manufacture products. The fast-growing number of personal printers, which are available for fair prices, and sold either as fabbers (do-it-yourself kit, DIY) like "Prusa Mendel" (Table 3.2, left) or as desktop printers (fully equipped machines) like "Mojo" (Table 3.2, right) is significant. Currently (2018), more than 500 machines are available in the price segment up to €5000 (or US$5000).

Table 3.2 AM Machines: Category Personal Printers (Fabbers and Desktop Printers)

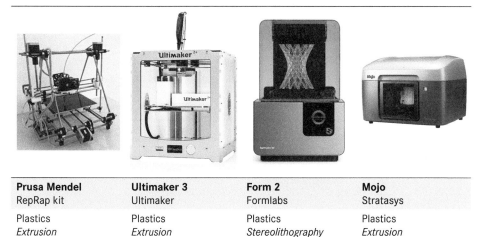

Prusa Mendel	Ultimaker 3	Form 2	Mojo
RepRap kit	Ultimaker	Formlabs	Stratasys
Plastics	Plastics	Plastics	Plastics
Extrusion	*Extrusion*	*Stereolithography*	*Extrusion*

3.2.1.1 Fabber and Do-It-Yourself Printers (DIY)

Fabbers are personal printers, mainly used privately, which in most cases are self-built or -assembled (Figure 3.7). The term *fabber* is an abbreviation of *fabricator* and refers to the background philosophy that anyone is able to manufacture anything with these machines. Web-based "blogs" and "fabber communities" are established to exchange information about building parts and carrying out possible modifications (pimping) of fabbers, and to propagate new forms of cooperation like "cloud fabbing". Users develop their skills via mutual training or as autodidacts. All that is required is computer literacy and the kitchen table to operate the device.

In combination with a personal computer (PC), fabbers become complete manufacturing units for an individual and decentralized production.

 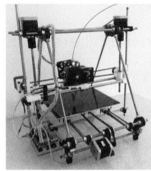

Figure 3.7 DIY fabber "Prusa Mendel" kit: illustration taken from the construction manual (left); assembled fabber, ready for operation (right)
(Source: RepRap/GoetheLab, University of Applied Sciences, Aachen)

Kits for fabbers can be ordered via the internet commencing at €400, inclusive of corresponding assembly instructions. They can be assembled without previous special knowledge; potential users should only not be shy to use a screwdriver.

Most of the fabbers on the market today (2018) operate according to the FLM process: heating of a prefabricated filament (made of thermoplastic) using a nozzle and depositing of the material on the build platform (or on a previously generated part) string by string to form a layer, and layer by layer to form the part. Today, filaments are available from different plastics and plastic blends, mostly from PLA and ABS. Rolled up on a reel, the filaments are available with standard diameters of 1.75 or 3 (2.85) mm and are offered in different colors.

With the introduction of fabbers, a years-old dream became a reality: children design their toys or the elements of their construction kits and manufacture them immediately by themselves. 3D printing "blogs" offer help and support the exchange of experiences. For this kind of application, productivity in many cases is of minor importance; partly manual work and repeated prints are accepted. Figure 3.8 shows parts that were printed by a fabber. Figure 3.8 shows exactly matching supports with bearings installed after the print (left), and a cut-away turbine blade with internal cooling channels; the part is made from plastics to serve as a demonstration object (right).

Figure 3.8 Individual and completed manufacture: fabber; support with installed bearings (left); cut-away picture of a turbine blade (right) (Source: IwF GmbH/GoetheLab, University of Applied Sciences, Aachen)

Table 3.3 Personal Printer: Fabbers Overview (Characteristics)

Personal Printers: "Fabbers" – DIY Printers	
Manufacturer	Communities
Price of machine	Kits from approx. €400 onwards
Software	Open source
Build materials	Plastics (ABS, PLA, nylon, PE, PVA, HIPS, blended materials…)
Advantages	▪ Low-cost machines
	▪ Low-cost materials
	▪ Large variety of materials
	▪ No requirement regarding infrastructure (plug only)
Disadvantages	▪ Operation of machine: no "plug and play"
	▪ High rate of disposals/scrap
	▪ In many cases support structures cannot be applied

Most of the personal printers available on the market work according to the FLM process. However, there is a trend to offer different processes like sintering or stereolithography in a "low-cost" version. The target customers are private individuals and small companies.

The Sintratec Company from Brugg, Switzerland, for example, has commercialized a kit for about €5000, which works according to the laser sintering principle and processes with PA 12 powder. Figure 3.9 (left) shows the machine, assembled from the kit, and parts with functional integration made from PA 12 (right).

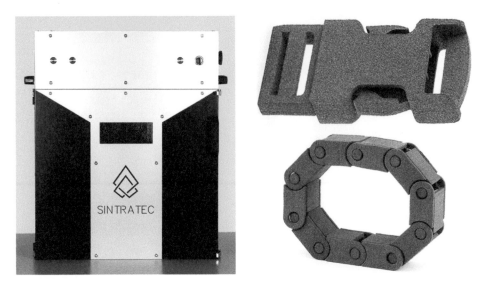

Figure 3.9 Fabber: Sintratec Kit – laser sintering (left); sintered functional prototypes, made
from PA 12 (right)
(Source: Sintratec)

Sintratec claims that the machine can be operated easily and can be integrated into
any production site.

The first limited series of 60 sets was financed by a crowd-funding campaign in a
month and was shipped half a year later. Thus, and with the help of numerous
supporters, the cornerstone of the Sintratec company was established.

3.2.1.2 Desktop Printers

Desktop printers are predominantly used in the professional or semi-professional
area. They are manufacturing units, placed in the lower price range, but not DIY
systems. After a short training, these easy-to-handle printers work with sufficient
accuracy.

Today (2018), desktop printers following the FLM process, as well as inexpensive
printers that generate parts using stereolithography or the laser-sintering process,
are available.

Software for desktop printers is either developed or bought by the manufacturer
and put to the client's disposal when purchasing the printer. A simple software
interface and preset process parameters (for example "low", "medium", and "high"
quality) allow non- (manufacturing) technicians to produce parts with suitable
properties.

Table 3.4 Personal Printer: Desktop Printer Overview (Characteristics)

Personal Printers: "Desktop Printers"	
Price of machine	From €1000 up to €10,000
Software	Manufacturer-specific, plug & play/open source
Build materials (depending on process/machine)	Plastics (ABS, PLA, nylon, PE, PVA, HIPS, blended materials, epoxy resin...)
Advantages (depending on process/machine)	▪ Low-cost machines ▪ Limited support structures ▪ Preset parameter sets ▪ Many options for accessories (washing machine for the removal of the supports, scanner, camera for process monitoring, integration into LAN/WLAN, etc.) ▪ No requirements regarding infrastructure (plug only)
Disadvantages (depending on machine)	▪ High material costs ▪ Depends on manufacturer of machine (software/material)

For most of the systems it is possible to generate individual parameter sets in an "expert mode", which allows the choice of the best part- and application-related setting.

Figure 3.10 (right) shows an approximately 45 cm high bust, manufactured by the desktop printer Makerbot Z18, shown in Figure 3.10 (left). It demonstrates the production of art and cultural heritage objects. The printing time of the bust was 72 hours with a layer thickness of 0.1 mm ("high quality").

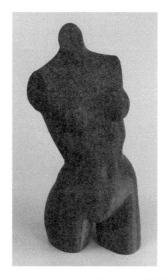

Figure 3.10 Individual manufacture of a finished part: desktop printer Makerbot Z18 (left); bust, printed by Makerbot Z18 (right)
(Sources: Makerbot Industries (left); GoetheLab, University of Applied Sciences, Aachen (right))

Desktop printers work self-sufficiently, and therefore automatic operation/print overnight or over several days should not cause problems. Moreover, many machines can be linked to the user's LAN/WLAN, which allows external monitoring, operation, and control of the printing process by an integrated camera. By distance monitoring, an observation of the printing process by an operator directly at the machine is not necessarily required.

Desktop printers are stand-alone machines and not integrated into a production system. Operational steps like change of material, removal of the completed part, and, if required, post-processing and finishing must be done by a human operator.

Desktop printers are offered that work according to different processes others than the FLM process and therefore can process other materials. Figure 3.11 (left) shows the *Form 1+* desktop printer of the Formlab company that works according to a modified stereolithography process. The build platform is moved layer by layer out of the build chamber/build tank, which is filled with liquid epoxy resin. To solidify the liquid epoxy resin, a laser outlines the contour of each single layer. This is done "from below" through a glass plate. Thus, the part is growing "upside down". Figure 3.11 (left) shows the printer and the printed part, "hanging" upside down from the build platform. In Figure 3.11 (right) the printed ornamental sphere is shown after separation from the build platform and removal of the supporting materials.

Figure 3.11 Individualized manufacturing of final parts: desktop printer Form 1+ (left); ornamental sphere, made from epoxy resin, printed by Form 1+ (right)
(Sources: Formlabs (left); GoetheLab, University of Applied Sciences, Aachen (right))

The subsequent machine *Form 2* processes layers with a thickness of 25 up to 100 microns. It works with a laser. The beam diameter is 140 μm and the laser power is 250 W.

The company SinterIt from Kraków, Poland was the first to develop a powder printer in desktop size with the aim to spread laser sintering to the wider public. The machine *LISA* was presented for the first time on the Hannover Fair 2016 and can be ordered internationally online.

LISA (Figure 3.12) works according to the laser sintering process and processes PA 12, with layer thicknesses up to 0.06 mm.

Figure 3.12 Individualized manufacturing of final parts: desktop printer LISA (left); "low-cost sintering" parts, made from PA 12 (right)
(Source: SinterIt)

3.2.2 Professional Printers

Professional 3D printers show the typical characteristics of AM processing centers. This requires an organized workshop. Programming is independent of the build process and increases the overall flexibility. Some manually executed program steps are already integrated and are processed (partly) with automation; this, for example, is the case for closed material cycles. The sequence of the program is completely automated. There are the rudiments of process monitoring. The operator gets feedback from the system.

Professional printers are stand-alone machines for the fabrication of functional parts (see Table 3.5). The main application of professional printers is the commercial use in the office or workshop. For most of the printers no special infrastructure is required, only a plug point and a table. A separate office room makes handling of material and parts easier and reduces the noise emission, which generally is not significant.

Table 3.6 provides an overview regarding professional printers with respect to machine prices and materials, and marks advantages and disadvantages of the machines.

Table 3.5 AM Machines: Category Professional Printers (Office Machines)

Agilista-3200W	Dimension Elite	Objet 30 prime	ZPrinter 450
Keyence	Stratasys	Stratasys	3D Systems
Plastics	Plastics	Plastics	Gypsum plaster
Polymer printing	*Extrusion*	*PolyJet*	*3D printing*

Table 3.6 Professional Printers: Overview (Characteristics)

Professional Printers	
Prices of machines	From approx. €20,000 up to approx. €70,000
Software	Manufacturer-specific, plug & play
Build Materials (depending on process/machine)	Plastics, ceramics, metals, gypsum-starch powder
Advantages (depending on process/machine)	▪ Preset parameter sets ▪ Minimal/no infrastructure required ▪ Short training ▪ Office infrastructure
Disadvantages (depending on process/machine)	▪ Partly high material costs ▪ In part, the user depends on the manufacturer of the machine (software/material)

3.2.3 Production Printers

Production 3D printers (see Table 3.7 and Table 3.8) show the characteristics of flexible additive manufacturing units. They may come as stand-alone machines but increasingly they are designed as AM processing centers, which are equipped even with integrated automated peripheral devices, like de-powdering stations. Characteristics are additional features that enable a detailed operational planning like nesting on the job, precise build-time estimators, or even simulators of the manufacturing process. The common target of production printers is an operation with a minimum of manual process steps, a mixed operation regarding the parts, continuous manufacturing, and production change free of preparation, etc. All these are characteristics that are regarded as advancements on conventional subtractive manufacturing, and can be considered as system-inherent with regard to AM. Current examples are the production printers of SLM Solutions (see Figure 3.13) and EOS (see Figure 3.14).

Table 3.7 AM Machines: Category Production Printers

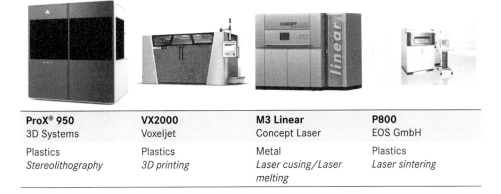

ProX® 950	**VX2000**	**M3 Linear**	**P800**
3D Systems	Voxeljet	Concept Laser	EOS GmbH
Plastics	Plastics	Metal	Plastics
Stereolithography	*3D printing*	*Laser cusing/Laser melting*	*Laser sintering*

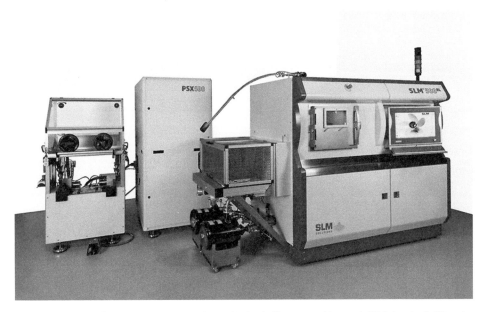

Figure 3.13 SLM® 500 HL: Laser melting units including unpacking unit PRS (to the left) and powder screening station (middle)
(Source: SLM Solutions)

Figure 3.13 shows the production printer SLM® 500 HL of the SLM Solutions company, a laser melting unit that processes metals. Regarding the machines of SLM Solutions different processes of layer generation have to be distinguished. In the standard version the layers are built according the basic laser melting process. The larger machines additionally offer a version with the so-called "double beam" technology, which uses two different lasers. One laser operates with low power and a small focus, generating the edges of the layers. The other one operates with high power and a top-hat energy distribution (profile), which is used to build the inner areas inside of a contour. Because of the split of the lasers in order to process either the contour or its inner area, this process of SLM Solutions is called shell-core strategy. This name has been generally adopted.

Table 3.8 Production Printers: Overview (Characteristics)

Production Printers	
Prices of machines	From approx. €130,000 up to approx. €1,800,000 and more
Software	Manufacturer-specific, plug & play
Build material (depending on process/machine)	Plastics, ceramics, metals, gypsum–starch powder
Advantages (depending on process/machine)	▪ "Ready to print" parameter sets ▪ Large varieties of printing material ▪ Minimum rejects
Disadvantages (depending on process/machine)	▪ Post-processing required ▪ High operation costs ▪ High material costs ▪ Workshop infrastructure required

The SLM® 500 HL is designed as production machine and as production system. It is equipped with a build space up to a maximum size of 500 × 280 × 325 mm and works with four synchronized YLR fiber lasers.

SLM Solutions offers a powder screening station (PSX) designed for the SLM® 500. It helps to empty the powder bin automatically. It can be refilled under an inert gas atmosphere. The powder screening station separates reusable powder from waste and recycles the powder continuously via intermediate storage in a bin. Peripheral equipment enables an extension to a (partly manual) production system. An "unpacking unit" (PRS) enables a quasi-continuous (parallel) operation by means of interchangeable additional build cylinders.

Another example of a production printer is the EOS P 396 (Figure 3.14). The EOS P machines work according to the laser-scanner principle and process plastics. It uses a 70-W CO_2 laser. A "dual focus" module is also available. Its beam diameter can be varied during the build. Thus, the contour can be generated very precisely, but slowly, while the areas inside are processed using a larger beam diameter, which is thus significantly faster.

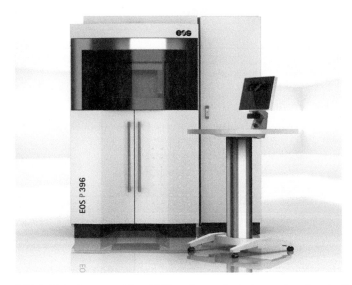

Figure 3.14 EOS P 396: sinter machine (SLS) and manufacturing system for plastics (Source: EOS GmbH)

3.2.4 Industrial Printers

Further developments lead to "flexible AM systems" (FAS). Manufacturers concentrate on metal printing according to the SLM process and typical are elements of automation technology. These include storage of build platforms, multi-machine (= printer) operation, automated manufacturing control, supervision by computer and DNC (distributed numerical control), interlinking of the part's production steps and transportation equipment, as well as a closed material cycle. Process monitoring and control are of central importance. Thus, significant requirements of the "Industry 4.0" strategy are fulfilled with an AM system. The target of a untimed manufacture, which is important for non-additive manufacturing, is system-inherent regarding AM manufacture.

According to the details of FAMS published so far, the separation of the build and the handling sections as well as the separation of build chamber and exposure unit (regarding the metal-powder process: laser) are common characteristics. Also, the use of up to four lasers (multi-laser technology) and the size of the build platform of approximately 400 × 400 × 400 mm are, to a large extent, common features.

Table 3.9 Industrial Printers (Characteristics)

Industrial Printers	
Price of machine	Starting from €1,200,000; including modules up to €2,000,000
Software	Manufacturer-specific, plug & play
Materials (depending on process/machine	Metals (up to now)
Advantages (depending on process/machine)	▪ Modular production system with attached heat treatment unit and storage device ▪ Individually adjustable ▪ Increase of productivity
Disadvantages (depending on process/machine)	▪ Post-processing required ▪ Very high machine costs ▪ Large amount of material leads to high investment in materials ▪ Workshop environment required

Closed material systems allow an automated preparation of the build material and the removal of unutilized powder. Build platforms or pallet systems provide the basis for continuous manufacturing and the required intralogistics.

MetalFAB1 – Additive Industries (AI)

With the company Additive Industries (AI), Eindhoven, Netherlands (founded 2012), a new supplier entered the market initially with a study in 2014. In March 2016 the first β-machine was delivered to the Airbus production company APWorks GmbH, Taufkirchen, Germany.

The AI concept integrates all process steps in one line and transfers the part, including supports, on the build platform to final processing (post-processing and finishing).

The machine is a freely programmable monoblock unit, as shown in Figure 3.15. It consists of at least four modules. These are the central control module "Controls", which also contains the filters; the 3D printing module "AM Core"; the included exposure module ("Exposure"), where laser and scanner are installed; and a cleaning and discharge module ("Exchange"). This constellation is called "3 + 1" by Additive Industries.

Figure 3.15 Flexible AM system, Additive Industries, MetalFab1: flexible AM systems (FAMS) for series production of metal parts. Development stage: control module, exposure module and two process chambers, heat treatment module, storage module, and discharge module
(Source: Additive Industries)

With a further print module, which increases the productivity, a storage module (storage) for the build platforms, and a heat treatment module, the machine can be extended to six modules (this version is installed at Airbus APWorks GmbH). Two print modules can be served by one exposure unit.

At the rear side the modules are connected by a robot on tracks that shifts the build platforms between the single stations.

It can be assumed that the price for the "fit-for-use" minimum configuration will be between 1.2 and 1.5 million Euros and unit with six modules will cost about 2 million Euros.

Factory of Tomorrow – Concept Laser/GE

Concept Laser/GE has presented the study "factory of tomorrow". The modular assembly of the system is realized by freely combinable elements like the handling station and the process station. Two build modules allow the simultaneous preparation and processing of the build job and thus an uninterrupted operation is possible.

The elements material storage and supply cylinder, overflow cylinder, and build module (with separate exposure unit) are integrated flexibly into the production process by a driverless transport system. Tunnels between the back-to-back arranged (build- and powder-) modules guarantee safe transport routes between the stations. There is an automated de-powdering and screening station. By means of these modules fully automatically working, digital factories can be assembled (Figure 3.16).

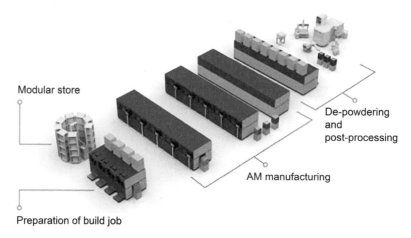

Figure 3.16 Study "Factory of Tomorrow" (Source: Concept Laser/GE)

SLM 800 – SLM SOLUTIONS

SLM Solutions, Lübeck, Germany presented an industrial-grade *Flexible AM System* (FAMS) called SLM 800 at the Formnext Fair 2017 in Frankfurt, Germany.

The SLM 800 machine and system is designed to be an expandable, integrated solution. Its core component is a fully automated handling station that automatically manages processes like pre-heating, cooling down, unpacking, and powder removal, including parts and powder transport in a sealed powder cycle with a permanent filtering process. Thus, manual handling is totally avoided, especially beneficial when making large parts. However, using the SLM 800 system or other integrated (metal) systems of today (2018), post-processing and finishing must be done separately.

The SLM 800 system is not only designed to manufacture large industrial parts, but also to improve productivity, as it can be extended up to six printers, controlled by just one handling station (Figure 3.17) and thus can be integrated into an "Industry 4.0" environment.

Figure 3.17 Industrial-grade flexible AM system (FAMS) SLM 800: fully automated handling station (left); one of up to six integrated printers of the SLM 500 type (on the platform) (right)
(Source: SLM Solutions)

The printers can work inside an outline of 500 × 280 × 850 mm. This equals the standard dimensions (500 × 280 × 365 mm) of the SLM 500 with an enlarged z-axis. The printers can be equipped with up to four 700 W lasers. New quality features are melt pool monitoring (MPM) und laser power monitoring (LPM) in conjunction with an optimized machine control software (MCS) and the already mentioned vacuum supported closed powder cycle (PSV). Besides hardware, the company also demonstrated a new software solution named "Additive Designer", a pre-processing software tool that combines all features necessary to generate an AM part from imported CAD data to define milling strategies for post-processing.

While the teams involved in metal printing are still working intensively on a fully automated 3D printing process, the developers of plastic printing systems seem to be ahead. 3D Systems already launched its structurable, modular "Figure 4" production system in late 2014, but it took until the end of 2017 until the first shipment to an unknown "Fortune 50" customer was reported.

"Figure 4" – 3D Systems

The "Figure 4" concept is a combination of SLA 3D printing technology, robots, and other handling elements that pass a 3D print through a number of processing stations that mimic the classical SLA process (SLA: Stereolithography Apparatus, a brand term of 3D Systems). After the part is printed, it is automatically moved to a washing station, a post-curing station, a painting station, or other finishing or manufacturing steps required for the particular part (see Figure 3.18). 3D Systems does not offer all imaginable options yet but will implement and commercialize by customer request.

UI   4 Print Engine
Module   Module

Figure 3.18 Industrial-grade flexible AM system (FAMS) "Figure 4", 3D Systems: fully configurable modular, and automated production system controlled by a user interface (UI) control unit, suitable for up to four SLA printing modules (Source: 3D Systems)

Standard configurations are:

- Production unit:

  completely automated and customizable system for high-volume production, with integrated post-processing.

- Modular unit:

  expandable system, to grow with the user's needs.

- Stand-alone unit:

  industrial-grade, single engine 3D printing system, suitable for small business.

- Dental unit:

  independent 3D printer, designed for dental production.

Professional printers are able to rapidly produce prototypes and final parts for many industries, including consumer goods, automotive, aerospace, and medical. The printers themselves as well as the materials are literally taken from the (3D Systems) shelf. Due to the complexity of the system, the "Figure 4" modular printer platform enables commercial enterprises to economically produce thousands (or more) of products per year. 3D Systems only ask for just one basic decision: industrial application or dental application.

The price depends on the configuration, therefore being entirely variable. The basic price of US$25,000 (*www.aniwaa.com/product/3d-printers/3d-systems-figure-4/*) must be regarded as a "starter price" rather than a lump sum for a FAMS. The system therefore is more like a toolkit than a product. As the system is designed for additive mass production, the price is less important and will be over-compensated by efficiency advantages. The costs per part could drop by up to 71% due to the reduced labor. According to 3D Systems, parts are produced up to 14 times faster and with 1.5 times less waste. In the end, labor costs are reduced by 4 times.

◼ 3.3 Conclusions and Outlook

3D Printers cover all possible applications. They are available as fabbers, office machines, or machines for direct production from numerous suppliers.

Personal 3D printers marked the start of a new AM era and established a new market for "low-cost" machines. It also demonstrates the entry of 3D printing into the mass market.

In future different variations of personal printers will enable everybody to manufacture parts by themselves and share this ability with a huge community of users worldwide. The members of this community are designers and manufacturers. They share their own products in all conceivable ways. The digital revolution has started to become three dimensional!

In addition, this development leads to flexible AM systems (FAMS). Characteristics of the first machines are a high degree of automation and the complete integration of the entire production process including process monitoring, all this aiming to fulfill all significant requirements of the "Industry 4.0" strategy.

◼ 3.4 Questions

1. **Where do the 3D data sets for manufacturing come from?**

   In the engineering department data sets in most cases are obtained from 3D CAD design and are exported in neutral data formats (e.g., STEP) or as STL (or AMF) format directly from the CAD data sets. For medical applications most of the data sets are generated from CT (or MRI) scans of patients.

   The industry increasingly uses (specialized) CT scanners, because these provide non-destructive total recordings of the data and allow a direct (false-color) quality check with the original 3D CAD data set.

2. **What is the STL format and what does it define?**

   The STL format is the standard format to transfer data for additive manufacturing. To gain a STL data set for a part, the free-form surface (inner and outer) is approximated by triangles. STL is a very simple data format. Each triangle is defined by the coordinates of only three corner points (x, y, z-parameters) and a normal vector, defining the orientation of the surface.

3. **How do the formats STL and AMF differ from each other? What is the advantage of AMF?**

In the "Additive Manufacturing File" (AMF) format, besides geometry also information concerning different materials, colors, textures, and physical-technological properties are stored.

By means of the AMF format, machines can be addressed that generate textures, colors, multi-material and gradient material, as well as micro structures directly in the process, which is not possible with the STL format.

4. **What does "errors in the STL data set" mean?**

STL data sets derived from a CAD system may contain errors that cannot be attributed to careless working with the CAD system or during scanning. During data exchange between different CAD systems, errors may occur. Meanwhile, good repair software is available; therefore, errors are not a major issue anymore.

The most frequently occurring errors in STL data sets are holes or gaps, overlapping triangles, and wrong normal vectors (see Figure 3.5).

5. **What is the difference between the process chains rapid prototyping and rapid manufacturing? Characterize both process chains.**

Regarding the process chain rapid prototyping the design is developed based on the rules of series design, whereas the process chain rapid manufacturing is based on the design rules of AM.

The choice of material regarding rapid prototyping is decided by the AM manufacturer, to simulate the series material. In both process chains the choice of parameters for the build process is decided by the AM design in accordance with AM process parameters. For rapid manufacturing in addition the instructions of the AM design are to be observed.

Detailed descriptions of the process chains are to be found in Section 3.1.1.1 "Process Chain: Rapid Prototyping" and Section 3.1.1.2 "Process Chain: Rapid Manufacturing".

6. **How can 3D printers be classified?**

Personal printers with the sub-categories "fabber" and "desktop printer"; professional printers, production printers, and industrial printers (see Figure 3.6 and Section 3.2 "Machines for Additive Manufacturing").

7. **What is the main difference between "fabbers" and "desktop printers"?**

"Fabbers" are personal printers, used predominantly in private surroundings, which in most cases are installed or assembled by the users themselves. The installation, design, and software to operate the printer are mainly open source.

Desktop printers are predominantly used in private or semi-professional surroundings and are manufacturing machines at a lower price level, but not DIY systems. "Desktop printers" are printers with sufficient accuracy, which are easy to operate after a short training.

8. **How can "professional printers" be distinguished from "production printers"?**

"Professional printers" show typical characteristics of AM processing centers, including a workshop organization. The process is completely automated; a few manual steps are integrated, but are (partly) automated. There are approaches for process monitoring and the production of functional parts is the main intention.

"Production printers" are stand-alone printers with peripherally arranged and linked automated equipment, with the aim to produce final products.

9. **How can "production printers" be distinguished from "industrial printers"?**

Production printers deliver one or a small number of final parts. Industrial printers are "Flexible AM Systems" (FAMS) that continually produce a large number of final parts under productivity boundaries. Characteristics are automated devices, like build platform handling and storage, multi-machine operation, automated production control, interlinking of the part's production steps and transportation equipment, and a closed material cycle. Process monitoring and control are of central importance.

References

[1] Gebhardt, A.; Hötter, J.-St.: *Additive Manufacturing: 3D Printing for Prototyping and Manufacturing,* Carl Hanser Verlag, Munich, 2016

4 Applications of Additive Manufacturing

This chapter deals with the additive manufacturing of parts and their application. First, the basic application levels according to Figure 1.4 are considered, and the question of whether we are talking about rapid prototyping or about rapid manufacturing is raised. What makes a part a prototype, and what distinguishes a prototype from a product? Surprisingly, neither the material nor the additive manufacturing process nor the machine determines whether a part is a prototype or a product. The central question concerns how the requirements of the design are met by the part.

The core of this chapter discusses the application of additive manufacturing. Accordingly, it is subdivided into application sectors or branches. The examples underline the fact that no fixed relation exists between a specific application or branch and a special additive manufacturing process. In reality, alternative applicable processes exist that often are even competing against each other. Depending on the situation, indirect or secondary AM processes have to be added to achieve the desired results.

The choice of the examples cannot be considered to be exhaustive or mutually exclusive, as it probably represents neither the optimal nor the only possible solution. As nearly every industrial branch uses additive manufacturing, each single sector of industry cannot be addressed. The discussion of the examples of this chapter can be considered as logical approaches to the applications of additive manufacturing. A more systematic approach with regards to strategy and particular design characteristics that are part of additive manufacturing processes is discussed in Chapter 5 "Perspectives and Strategies of Additive Manufacturing".

Regarding the application of additive manufacturing processes, industrial issues are highlighted. The majority of applications today are to be found in the field of product development of new serial parts and thus in the field of rapid prototyping. The portion of final products with regard to the application of rapid manufacturing is steadily increasing.

Apart from users with large design departments and high production volume, there is a steadily growing number of users and industrial branches applying additive manufacturing for the realization of new ideas and thus for the extension of the scope of their products. They often achieve impressive applications and parts, and can be considered to be the true motivating force behind new solutions and ideas. In this section an attempt is made to present the success of the industry as well as new aspects to promote the dissemination of additive technology. Today most industry sectors use AM; the presented examples represent a small selection of what is possible.

■ 4.1 Automotive Industry and Sub-Suppliers

Automobile manufacturers and their sub-suppliers have been among the first adopters of AM, after the technology emerged in the 1980s. Moreover, they have been the pioneers in use of 3D CAD systems since inception. Therefore, 3D design has not been a hurdle for the automotive industry in applying AM. This is the reason why generating the required perfect 3D data sets has never been an issue for automobile manufacturers and their sub-suppliers.

As the number of prototypes increases due to diversification, product variations gain importance. Prototypes are used by companies for internal validation, and form fit and function. They are an excellent marketing tool for discussions with customers and suppliers. Customized products mitigate direct manufacturing, significantly reducing tool costs.

4.1.1 Automobile – Interior Components

The design of the interior of an automobile determines the character of a vehicle and often influences the buying decision. In contrast to the exterior design, the interior normally is formed from various single parts that originate from numerous sub-suppliers. The majority of parts are manufactured by injection molding. Therefore, additively manufactured parts are primarily used for tests and presentations of car concepts.

But in the future, AM parts will be increasingly installed directly in cars, mainly because of the decreasing size of series and increasing number of variations. Although all additive manufacturing processes can be applied for the production of interior parts, the preferred processes are laser sintering, extrusion, and polymerization. Usually, laser sintering and extrusion lead to directly applicable parts, whereas by means of stereolithography or polymer printing typically master models for secondary processes are generated.

In Figure 4.1 (left), a dashboard insert for a radio or the control elements of the air conditioning is displayed. As the geometry depends on the car's equipment, finally determined by the buyers taste and budget, many different design variations are required.

The fuel tank shown in Figure 4.1 (right) is a very complex technical part that is installed inside of the vehicle, and not visible to the customer. Often the fuel tank has to be fit into the available space that is provided in the first design concept. Sensors for fuel level control are integrated into the tank. As the tank is a large part to produce by AM, only a few machines – for example, the EOS P 800 – are able to produce it in one piece. Alternatively, parts to be produced by means of laser sintering can be divided with the help of suitable software, manufactured in sections, and assembled by gluing. Sometimes this procedure can be even advantageous with regards to an economical use of the available installation space or to prevent distortions. The tank can be sealed and tested as a 1:1 scale prototype. Laser sintering of polyamide provides stress-resistant parts like the dashboard insert or the fuel tank, but results in a reduced surface quality compared to the polymerization processes.

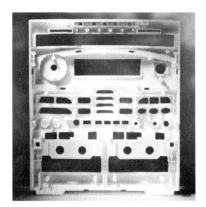 

Figure 4.1 Dashboard insert: laser sintering, polyamide (left); fuel tank: laser sintering, polyamide (right); both parts 1:1 scale
(Source: EOS GmbH)

In contrast, the speaker housing depicted in Figure 4.2 (right) shows a very good surface quality and therefore it can be used like a series part. It was manufactured as a RTV copy from a polished stereolithography master model. The AM master model provides the detailed geometry and the surface quality while the colored and loadable material and the desired number of parts are produced by the RTV process (Figure 4.2, left).

Figure 4.2 Speaker housing: stereolithography, master model and silicone mold (left); casted
part, RTV (right)
(Source: CP-GmbH)

Interior lamps, as well as front and rear lamps in particular, are frequent objects of
design variations and face-lifts. Design variations can be generated quickly with
high quality by additive manufacturing and then be evaluated (Figure 4.3). Transparent elements are important features of the lamps.

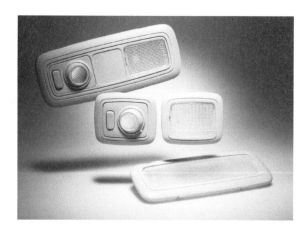

Figure 4.3 Interior lighting: design variations; stereolithography and RTV, including inserts
(Source: CP GmbH)

The elements are produced additively as non-transparent parts. Afterwards they
are cast as transparent and colored (if needed) parts by means of RTV. Even the
textures of the diffusion pane can be achieved by special foils, inserted into the
silicone mold. Elements such as switches or frames for lenses are manufactured
separately by the same procedure and inserted into the final part (Figure 4.3).

Like many other parts, lamp sockets and covers produced by additive manufacturing are also used as spare parts for vintage cars in small series or as individually manufactured parts, even for racing.

4.1.2 Automobile – Exterior Components

Special editions of vehicles derived from volume car productions are often not only equipped with more powerful engines, but demonstrate their exclusive status by modified front and rear spoilers or side skirts, etc. As an example, Figure 4.4 shows a front spoiler for a very small special series. For economic reasons, manufacture by means of steel tools cannot be justified.

For this small series the spoiler was split into three separate sections (left, center, right). Each of the sections was manufactured by AM (laser sintering or stereolithography), finished, and assembled to a single master model with the help of a gauge. In a subsequent RTV process, a small series of spoilers was produced from polyurethane (PUR), varnished in the car's original color, and completed using the original decorative parts (grille, logo, trim) from the shelf of the vehicle's manufacturer.

Figure 4.4 Modified front spoiler for a vehicle's special edition; laser stereolithography, RTV, finishing, varnishing
(Source: CP-GmbH)

Metal elements for the power train or engine components can be manufactured in final product quality by means of additive manufacturing processes, primarily by selective laser melting (SLM). The parts can be single test parts or small series parts, to be used, for example, in racing cars. As an example, an exhaust manifold of a racing car used in a European car competition (Formula Student) is show in Figure 4.5.

Figure 4.5 Exhaust manifold of a racing car
(Source: Concept Laser/CE/TUfast e. V.)

The further development of additive manufacturing processes not only points to faster printing and shortening of the processing time. There is also a strong trend toward making large parts. The company "Local Motors", for example, presented the "Strati" (see Figures 2.31 and 2.32) as the first 3D printed car (2014) world-wide, printed on the BAAM (see Section 2.1.6.3 "Extruding and Milling – Big Area Additive Manufacturing (BAAM)", followed by the 3D printed bus "Olli" (see Figure 4.6).

Figure 4.6 "Olli": AM-based autonomous driving electric city bus
(Source: Local Motors)

Essential parts of "Olli" were printed by the BAAM extrusion machine with surface finish by milling (see Figure 2.32).

■ 4.2 Aerospace Industry

Due to small series and particularly to the requirements of lightweight design, the aerospace industry tries to avoid tools and focuses on tool-less additive manufacturing.

A milestone with regards to the direct production of interior aircraft parts was the introduction of the flame retardant and low-smoke FDM material "Ultem".

Many parts used in the interior of aircrafts basically do not differ from automotive ones. Therefore, sample parts for automotive applications also can be considered as sample parts for the aerospace application.

The development of metal processes, as well as processes for processing of ceramic materials, allows the direct manufacturing of technical parts for the airframe as well as for the engines. Figure 4.7 shows a hot-air duct that was manufactured by means of 3D printing (metal) and subsequent heat treatment.

Figure 4.7 Hot-air duct for an aviation turbine: 3D printing (metal), post-processing
(Source: FHG-IFAM/Airbus/ExOne)

Figure 4.8 shows a combustion chamber element made by selective laser melting (SLM). It verifies the applicability of ready-to-use, complex 3D printed elements.

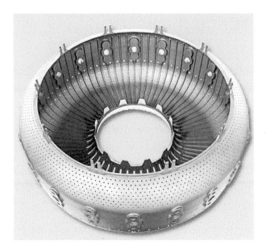

Figure 4.8 Combustion chamber element: selective laser melting (SLM)
(Source: Concept Laser/GE)

Directly printed parts made from metal are successful from technological and economic perspectives if they are optimized with regards to geometry and function.
AM offers the opportunity to apply a bionic design strategy, thus raising the potential for lightweight design (see Figure 4.9).

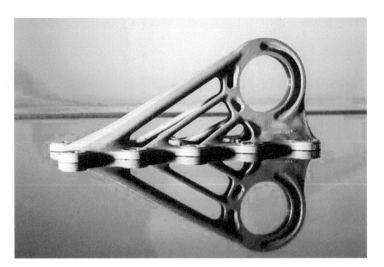

Figure 4.9 3D printed lightweight bracket, for the A350 XWB; SLM, titanium
(Source: Airbus)

It can be seen immediately that the parts look different from traditionally designed
ones. Thus, they show that for a fast and effective execution of AM an AM-suitable
design is mandatory; see Section 6.2 "Construction – Engineering Design".

■ 4.3 Consumer Goods

Consumer goods not only have to fulfill the intended functions, but also have to follow design trends. Therefore, they have to meet the sometimes rapidly changing needs and taste of consumer groups and their design preferences.

Lifestyle products serve a fast-developing market. As they undergo quick changes, trends have to be investigated, and marketing studies carried out, prior to the commencement of production. This requires prototypes. The cocktail cup shown in Figure 4.10 was manufactured for example by laser stereolithography, so its details and decorations can be reproduced. The cup itself was printed as a master model and cast by RTV from highly transparent material, while the base was only sprayed with silver paint, and the electrical equipment was installed.

Figure 4.10 Cocktail cup: laser stereolithography, RTV, and post-processing
(Source: Pfefferkorn/Toorank/CP-GmbH)

Bowles, vases, lamps, and other more decorative items are preferred objects for designers and serve as an experimental field to test the new freedom of design gained by additive manufacturing. Load-bearing parts are preferably manufactured by laser sintering of plastics (Figure 4.11, left), while complex, delicate, as well as transparent or translucent objects typically are produced by laser stereolithography or polymer printing (see Figure 4.11, right).

Figure 4.11 Lamp: laser sintering, polyamide (left); lamp: laser stereolithography, epoxy resin
(right)
(Sources: CP-GmbH (left); Freedom of Creation, FOC (right))

The three-dimensional representation of geodesy data is about to form an interesting niche market. For the production of corresponding models, the 3D printing
process, in this case the powder-binder process – e.g. Color Jet Printing (CJP) from
3D Systems – is of advantage, as the parts can be colored continuously, and the
manual coloring is dropped. Examples are contained in Section 4.8 "Architecture
and Landscaping".

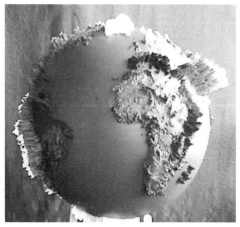

Figure 4.12 Relief globe: representation of the topography of the mountain heights (left) and
the ocean depths (right), each 250 times superelevated; laser stereolithography
and post-processing
(Source: Worldglobus)

Figure 4.12 shows two models of a globe. The models represent the topography of the mountains and the ocean floors. For better distinction of details, the mountain heights and ocean depths are superelevated in scale. The scale and the superelevation can be adapted individually for each model according to the needs of the user. To reproduce the fine spike-shaped details on the surface, stereolithography was used, requiring manual coloring.

Many other globes can be found in the internet, e.g. at "Thingiverse" or "Shapeways".

Mainly in the field of architectural design, increasingly fine and delicate structures are the trend, leading to ambitious requirements with regards to architecture models. For the manufacture of loadable, slim structures – as for example the roof structure of a stadium (Figure 4.13) – the plastic laser sintering process is established.

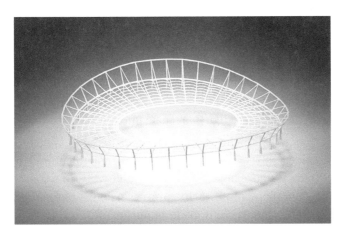

Figure 4.13 Roof construction of a stadium: laser sintering
          (Source: CP-GmbH)

The closer a product is linked to the anatomical characteristics of the users, the higher the degree of individualization has to be. The three-unit dental bridge in Figure 1.12 can only be used by one individual. Tool-bound manufacturing cannot be considered for these kinds of parts. Beside hand crafting, the only option is to use additive manufacturing processes. AM not only allows manufacture with the required complexity, but also production of a large number of differing parts within the same build process.

Further examples for individualized products, which not only depend on the anatomy of customers, but also on their taste, are individually styled glasses (Figure 4.14).

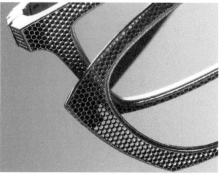

Figure 4.14 Customized frames for glasses: laser sintering, titanium
(Source: Hoet)

For example, the Hoya company presented at Opti 2017 – the international trade show for optics and design in Munich – an extensive program of customized spectacles frames. The system "Yuniku" also includes a scanner that records faces, facilitating the choice of suitable glasses and preparing the additive manufacture of the individual frame. The program was developed in cooperation with designers as Hoet (Brugge, Belgium) and Aoyama (Villeneuve d'Ascq, France) with technical support by Materialise. This approach was further developed into an internet-based tool, accessible to customers, which integrates the individual data into the 3D-printable frame design.

The extravagant sandals in Figure 4.15 can be considered as archetypes of individually designed, trendy shoes, which moreover can be manufactured in any size and height – as desired. As these objects can be used directly, they can be considered as products and thus the manufacturing process belongs to *rapid manufacturing*. Although the sample sandals were made by means of laser sintering from polyamide, extrusion and polymerization processes can also be applied with the correspondingly appropriate materials, including soft types like TPU and TPE.

Figure 4.15 "Paris Sandals", high heels: laser sintering, polyamide
(Source: Freedomofcreation, FOC)

The application of additive processes for the production of musical instruments was already presented in 2012 by Professor Olaf Diegel at the Massey University in Auckland, New Zealand with an example of a laser-sintered electric guitar made from polyamide.

In May 2016 the first 3D printed guitar made from aluminum was presented. The complete body of the guitar was manufactured in one piece on the machine EOS M400 (Source: Xilloc; see Figure 4.16).

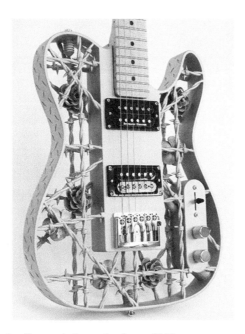

Figure 4.16 3D printed guitar made from aluminum, SLM
(Source: Xilloc / Diegel)

Meanwhile different service bureaus offer customized 3D prints from individuals. Full-body scanners with up to 80 digital cameras arranged inside a tube at different angles instantly acquire a 3D photographic image and thus a 3D data set. The advantage of this kind of scanner is that the recorded person does not have to keep still. It can thus be used for children, animals, and moving individuals. The up to 80 photos taken are loaded into a software application and automatically assembled to form a solid body. However, an additional manual post-processing needs to be applied and requires extended know-how in the field of image processing. The data set usually is printed by means of 3D printing (powder-binder process); see also Section 2.1.4 "Powder-Binder Process" and Figure 4.17. The post-processing of the mostly fine, delicate structures is based on infiltration but, depending on the desired quality, is laborious and time consuming. Alternatively, polymer jetting can be applied.

Figure 4.17 Body scanning and AM by means of 3D printing: powder binder process
(Source: Nathalie Richards (left); Caters New Agency (right))

Furthermore, the creative work of artists can be successfully supported by additive manufacturing processes. Sculptors often use clay models to build up and manually improve their work stepwise before final casting. Alternatively, a handmade model can first be scanned and transformed into an AM part from polyamide (by sintering) or plaster (by 3D printing). This master model can be improved manually, to reflect the "handwriting" of the artist, and then be transferred via a wax model (obtained by RTV) into a series of unique specimens, cast from bronze.

■ 4.4 Toy Industry

Although toys are regarded as consumer goods, the toy industry usually is considered separately. The majority of this industry involves the series production of plastic parts for children's toys, but increasingly also individualized models, mainly of automobiles, aircraft, and railroads, are focusing on adult clients. These models require fine details and a sensitive scaling that considers small and large details differently. Depending on the scale, some AM processes are more suitable than others. Figure 4.18 shows a model of a toy steam engine in scale G (1:22.5) that, including its tender, is about four feet long. For the manufacture of this object the layer laminate process is a good option, because the material is inexpensive, and the details are not too small. For this kind of (showcase) model, physical properties, such as load-bearing capacity, and very fine details are less important than the overall appearance.

Figure 4.18 Model of a toy steam engine, scale G (1:22.5): laminated object manufacturing
(LOM), paper; post-processing by varnishing
(Source: CP-GmbH)

In the case of scale reduction, e.g. to the most popular standard HO (1:87) for toy
trains, models with fine details are more important. For the manufacture of a
model, stereolithography is most appropriate. In Figure 4.19 the toy steam engine
of Figure 4.18 is shown in scale 1:87. In the background, it is displayed after the
manufacturing process including cleaning, while in the foreground it can be seen
after post-processing by polishing, spray painting, and adding fitting parts.

Figure 4.19 Model of a toy steam engine, scale HO (1:87): stereolithography; object after
printing (background); object after post-processing by polishing, coloring and
adding fitting parts (foreground)
(Source: CP-GmbH)

Many model enthusiasts desire exclusive objects that are not available as kits. The
models often are delicate, require fine details, and are very difficult to manufacture
conventionally, especially if hand crafting is required and only limited space is
available at home. As an example, the 3D CAD data set of the model of the "Anticosti" (Figure 4.20) – a Canadian offshore supply vessel and former minelayer –
was generated based on photographs on a home PC. The data set was printed by
laser sintering or stereolithography, depending on the required details.

Figure 4.20 Model of the Canadian offshore supply vessel "Anticosti": laser sintering and stereolithography, manual varnishing after the print
(Source: *https://www.schiffsmodell-magazin.de/?s=Anticosti*)

■ 4.5 Art and History of Art

Artists have been among the first using the almost unlimited freedom of design offered by AM processes. The Californian artist Bathsheba (*https://bathsheba.com*) generates individual objects by 3D printing of metal powders. Using a special heat treatment after printing, she achieves exclusive surface effects. Her objects are final products and are sold as art objects via the internet.

Metal was chosen as material because it is regarded as more valuable than plastics, due to its weight. Figure 4.21 (left) shows the object after manufacturing by 3D printing. Figure 4.21 (right) underlines that the special appearance is obtained by the finishing procedure.

The manufacture of individual human sculptures can be carried out by additive processes based on 3D data obtained from body scanners. This was arranged and demonstrated live by the conceptual artist Karin Sander already in 2002 at an exhibition in the Staatsgalerie (state gallery) Stuttgart, Germany. The objects of the exhibition were generated by body scanning of visitors directly followed by on-site 3D printing. At the opening of the exhibition only empty shelves, the scanner, computers, and two 3D printers were to be seen. At the closing ceremony all shelves were filled with sculptures of a large number of visitors (scale 1:7). For this application, monochrome 3D printing is a suitable process, because it is fast and inexpensive, and details are reproduced sufficiently. Figure 4.22 shows a group of these sculptures.

Colored 3D printing takes more time but enhances the effect.

Figure 4.21 Art objects: 3D printing (metal); object after AM manufacture (left) and post-processing as well as finishing by surface treatment (right)
(Source: ExOne/Bathsheba)

Figure 4.22 Body scanning and AM by printing (3D Systems/Z-Corporation): Plaster
(Source: Karin Sander/GoetheLab, University of Applied Sciences, Aachen)

Today the manufacture of individual sculptures by scanning and additive manufacturing is a rapidly growing business model; see Section 4.3 "Consumer Goods". At the CHIO Aachen, the official equestrian show of the Federal Republic of Germany, special awards are given for exceptional performances. Figure 4.23 shows "Altis MidMesh", a trophy that portrays a horse of the famous stud "Mehlkopf". It was scanned, reconstructed, and scaled (approx. 12 inches high). The data was prepared for the SLM process and printed from stainless steel (1.4404). The horse still with supports is displayed in Figure 4.23 (left). It is displayed in Figure 4.23 (right) after removal of the supports and intensive manual polishing.

Figure 4.23 Horse sculpture made from stainless steel: selective laser melting (SLM). After manufacture by SLM, still with supports (left); post-processed and polished (right)
(Source: Otto Junker / GoetheLab, University of Applied Sciences, Aachen)

In June 2016 the French architect Dominique Perrault was appointed one of 56 members of the "Academie de Beaux-Arts de l'Institut de France". Following an ancient tradition, new members bring their own usually classically designed sword to the ceremony. Perrault designed a sword that resembled reinforcement steel and thus follows a new understanding of design, including new surface structures. It was built by SLM from stainless steel. To be able to process the sword in the machine, it was divided into five sections and joined afterwards by laser welding (Figure 4.24).

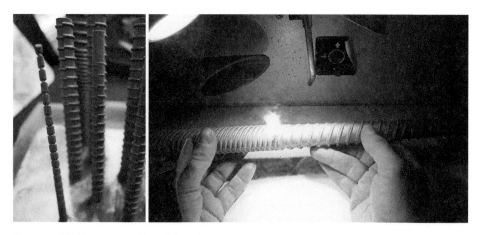

Figure 4.24 The sword of Dominique Perrault: view into the SLM machine after removal of the steel powder (left); joining the sections (right)
(Source: Otto Junker / GoetheLab, University of Applied Sciences, Aachen)

The tip was polished, and the handle was engraved (Figure 4.25).

Figure 4.25 The sword of Dominique Perrault: tip (above left); handle (above right); Dominique
Perrault presenting his sword in front of the "Institut de France" (below)
(Source: GKD Kufferath/GoetheLab, University of Applied Sciences, Aachen)

■ 4.6 Mold and Die Making (Rapid Tooling)

In this section, molds and mold inserts means molds for series production in the
sense of *direct tooling*. Molds for *prototype tooling* and *soft tooling*, as in the RTV
process, are discussed in Sections 1.2.2 "Indirect Processes" and 3.2 "Machines for
Additive Manufacturing".

Manufacturing of tools, molds, and dies basically just employs the negative of the
product data set. For the production of molds (negatives) the same additive manufacturing processes can be applied as for the manufacture of products (positives).

The additive manufacturing of tools, tool inserts, molds, and dies has been given its own name: *rapid tooling*. This was done primarily for marketing reasons: it was handled like a new technique to keep it topical. It is important to state that in applying rapid tooling no complete mold is generated, but primarily mold inserts, sliders, cooling sticks, or similar mold elements. As in the traditional mold manufacture, a mold generated by *direct tooling* consists of a combination of AM-based cavities and mold inserts, completed by standard elements.

The big advantage of additively manufactured series tools and tool inserts is the reduction of the cycle time due to enhanced cooling by conformal cooling channels. By layer technology, conformal cooling channels can be designed and manufactured close to the surfaces of the cavities. The channels do not have to be straight and do not have to have round cross sections, as is unavoidable with drilling.

Figure 4.26 shows the improvement of a mold insert by conformal cooling channels. The former drilled, straight channels, which frequently have to be closed by plugs, were replaced by conformal cooling channels that follow the freeform surface at a defined distance.

Figure 4.26 Conformal cooling: conventional design with drilled, straight cooling channels (above left) and by AM manufactured conformal cooling channels (above right); CFD (Computational Fluid Dynamics) simulation underlines the cooling effect (below left and right). Note: the temperature scales are not identical! (Source and details see ref. [1])

The part displayed in Figure 4.26 was made from tool steel by SLM and tested in the production process. As can be seen from the simulation of the production, which was confirmed in practice, optimized cooling channels have a considerable influence on the heat transfer and in this particular case led to a reduction of the cycle time by approximately 30%.

Another complex tool element with conformal cooling is shown in Figure 1.15. It is one half of a mold insert for the production of the upper part of a vacuum cleaner housing. The deep and narrow fins of the part, shown in Figure 1.15, have to be cooled down for a quick and safe demolding. After the design of the conformal cooling channels (dark), there is nearly no space left to drill the holes for the ejection pins. Therefore, air ducts (white) are provided, to carry out the ejection pneumatically. The AM tool half, made by selective laser melting, is shown at the right side of Figure 1.15.

A further development of the idea of conformal cooling results in complex channel systems or grids as shown in Figure 4.27. The CAD design is presented in the left part of Figure 4.27, and the cut-away AM part in the right part of Figure 4.27. To manufacture parts like this, selective laser melting is applied as preferred process. Using a suitable design, hollow parts can be built without supports and, due to the loose powder, be cleaned easily.

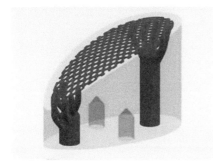 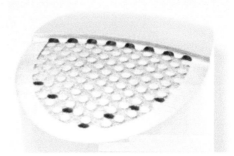

Figure 4.27 Cooling grid for conformal cooling: CAD design (left); cut-away view of the mold insert (right); selective laser melting, SLM (Source: Concept Laser/GE)

Suitable materials for tool manufacture – mainly tool steel, but also aluminum, titanium, and so-called superalloys such as Inconel or Hastelloy – are available for SLM.

■ 4.7 Medical Engineering

Humans are individuals and need individual medical treatment including individually designed (customized) aids such as implants, epitheses, or orthoses (leg braces). The required 3D data sets have to be generated by medical imaging processes such as computer tomography (CT) or ultrasound (US). A widespread data format for medical imaging and picture processing is DICOM (Digital Imaging and Communications in Medicine). Special software allows the conversion of the obtained grayscale images by suitable thresholding into material density charts, and a 3D reconstruction from the layer information. Thus, the basis is formed for a STL data set that can be used by any machine for additive manufacturing.

Laser stereolithography (Figure 4.28, left) and polymer printing are the preferred processes for the manufacture of medical models such as skulls and other human bone structures. This is mainly due to good surface quality and detailed reproduction. Interior hollow structures such as the orbital floor or the fine subcranial bone structure can be reproduced best by these processes. Also, by laser sintering, 3D printing, fused layer manufacturing (extrusion, FDM), or layer laminate manufacturing (LLM), medical models can be generated and therefore the mentioned processes are often used for this purpose. With regards to sintering and FDM there are particular authorized materials for medical applications that can be sterilized.

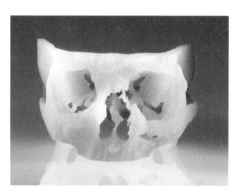 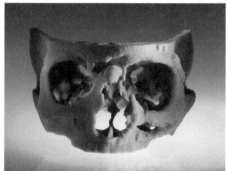

Figure 4.28 3D printed facsimiles of a human skull based on identical CT data: 3D printing
(left); stereolithography (right)
(Source: CP-GmbH)

Figure 4.28 (left and right) shows facsimiles of the facial bone of a human skull manufactured by 3D printing (powder-binder process) (left) and stereolithography (right). Both are made from the same data set. The differences based on the different processes are clearly recognizable. The powder-binder process delivers slightly fewer details, a coarser surface, and results in a non-transparent part. Whether these have to be considered as disadvantages or not depends on the intended use.

The decision for the process to be applied depends mainly on following issues:

- Will the part be implanted and for how long will it remain in the body?
- Will it be used only for training purposes?
- Does it need to be sterilized?

- Does it need to be transparent?
- What surface quality is desired?
- What is the required durability?
- How big is the available budget?

Figure 4.29 (left) shows an example of a medical application: a model of a human skull manufactured by laser stereolithography, and the adapted individual implant (Tailored Implant™), made from titanium. The implant was made by investment casting (lost wax casting) based on an additively manufactured master model. Therefore, the skull was scaled with respect to the estimated shrinkage. The fit of the implant was checked by means of a second stereolithography skull model in scale 1:1.

Alternatively, the implant can be designed by special software, e.g. Mimics of Materialise, and made by laser melting (SLM) or electron beam melting (EBM) directly from titanium or CoCr (see Figure 4.29, right).

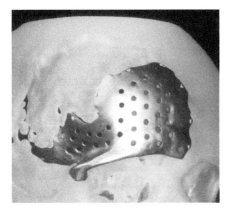 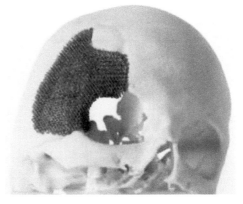

Figure 4.29 Skull models with individual implants: facsimile of a skull, manufactured by stereolithography (left) and by laser sintering (right). Individually manufactured implant from titanium, made by investment casting from a wax model that again was obtained from a stereolithography master (left) or directly by EBM (right) (Source: CP-GmbH/ARCAM/GE)

It might be advantageous to use an additively made lost wax master to cast the implant from titanium (Figure 4.29, left). In this case the wax model can be quality checked and, if needed, adapted by the medics prior to the final casting. In contrast, the directly sintered titanium implant must be modified via data set and printed again.

In general, 3D printing opens the chance to design more effective process chains by integrating manual process steps and 3D printed parts, thus freeing the specialists from routine work that could be done faster and more precisely by the 3D printer.

Often scanning is used to record even hidden structures. The scan data are reconstructed to form a 3D data model. The desired artificial organ then has to be adapted to the individual condition of the client. The necessary 3D modeling can be carried out by means of special software such as "Sensable" or "Geomagic®" haptic devices", and manufactured by AM.

Anaplastologists are skilled manual workers whose core competency is a deceptively real modeling of human body parts for prostheses. Before a patient obtains an applicable ear, a so-called "raw ear" made from wax is needed. It can be easily made by scanning and mirroring of the data of the healthy ear and be printed by means of the powder-binder process, followed by a wax infiltration (Figure 4.30, left). The part can be corrected afterwards in full detail by manually adding or removing wax. The final ear epithesis (Figure 4.30, right) results after counter casting with medical silicone material and a final decoration.

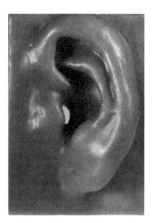 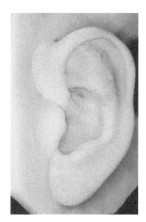

Figure 4.30 Ear epithesis: intermediate product (or "raw ear") after 3D printing by powder-binder process and wax infiltration (left); final ear made from silicone by counter casting of the wax master (right)
(Source: CP-GmbH/Professor Bier, Charité)

The same principle is applied by sculptors to achieve an elaborated wax master model for final casting.

White teeth and perfect tooth positions are important social indicators. Therefore, also adults wear dental braces, so-called "aligners". They are produced from transparent plastic film and therefore can also be carried over the teeth during daytime. The manufacture of the aligner is carried out by deep drawing of a transparent film over a teeth model. As the aligner has to represent the future tooth positions, and therefore cannot be taken from the existing teeth, a complicated process chain to predict the future alignment is designed.

The actual tooth positions are taken by the involved dentist using either a conventional dental impression or a scan generated by an intraoral scanner. In the case of a classical impression, it is sent to the company Align Technologies (*https://invisalign.com*) To achieve the required precision, the impression then is filled with a plaster-like white material, thus forming a positive of the teeth. Together with some other 10 positives it is embedded in a black contrast mass. The embedded impressions are milled in thin layers, and each single layer contour is scanned. From these scans a three-dimensional teeth model is reconstructed and transferred to a simulation program that allows depiction of all the patient's teeth virtually. With the help of this simulation, specialists determine the future tooth positions. As by means of an aligner only comparably small correction can be made per step, the total correction is split into single partial corrections. The results are transmitted to the involved medic for evaluation via the internet. From the data sets, all partial models required for the treatment are manufactured, each as a complete teeth model by means of stereolithography. On the build platform of a ProJet 7000 HD (3D Systems), for example, 90 teeth models are manufactured simultaneously.

As the teeth models represent the respective tooth positions during treatment, they are used to make the aligners as models for the thermoforming process. The contoured films are trimmed at the dental necks (Figure 4.31, left). The finalized aligners are handed over to the patient, who applies a pair of them, each of them for approximately two weeks, under medical supervision (Figure 4.31, right).

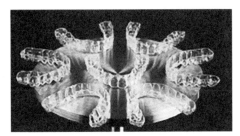

Figure 4.31 Thermoformed and trimmed aligner (left); set of finalized aligners (right) (Source: Invisible Align)

Another application of medical technique is shown in Figure 4.32: the hip socket was manufactures by SLM (approx. 10 hours manufacturing time). By individualization of the sphere geometry and the connecting points, as well as the inter-linked grid structure, the ingrowth of the implant is significantly improved by this design.

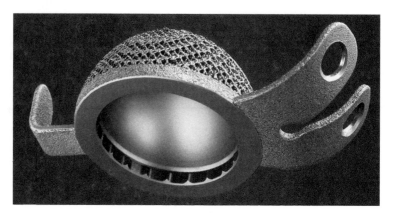

Figure 4.32 Individualized hip socket: SLM, titanium
(Source: Fraunhofer ILT)

■ 4.8 Architecture and Landscaping

Architects usually present their design by scaled models. Since they are working with 3D design programs, the required data sets for additive manufacturing are directly available. Therefore, the manufacture of models or model elements by additive manufacturing is comparably simple and spreads quickly.

Figure 4.33 shows a model of a mosque. The complex thin-walled freeform structure is very difficult to manufacture by means of traditional model making, but it is not a problem for AM. To get a detailed model, as well as a stressable part, e.g. to avoid damage as a consequence of touching, laser sintering was chosen. The cube-shaped elements were sintered, but also polymer printing would have been suitable. The model was primarily used for the public presentation of the project.

Another example is the display model of a planned tourist center shown in Figure 4.34. It is part of a concept that also includes a project description and a 3D presentation. For the same reasons as for the mosque, the model was built by laser sintering. Also, laser stereolithography or FDM could have been applied, if respective supports would have been provided. The powder-binder process would have caused problems, due to fine details, as for example the handrails. Polymer printing would deliver comparably good elements, but would need a large amount of support material, thus resulting in an expensive model.

Figure 4.33 Architectural model: mosque; laser sintering, polyamide
(Source: Deutschlandwoche)

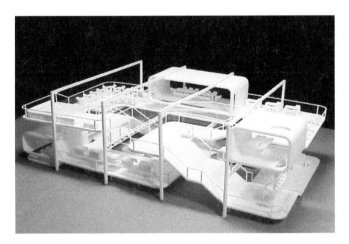

Figure 4.34 3D model of a concept of a tourist center: laser sintering, polyamide
(Source: Bernhard Bader)

Apart from the modeling of single buildings or parts of buildings, there is a growing demand for models of groups of buildings, villages, and entire landscapes. To highlight the advantages of a certain design or to emphasize landmarks, often coloring of the models is required. If no fine geometrical details have to be represented, colored 3D printing (powder-binder process) is the best choice (Figure 4.35).

Figure 4.35 Architecture models: two houses in rural surroundings
(Source: 3D Systems)

The data can be obtained from any type of geographic information system (GIS) or directly from the internet, e.g. from Google Earth or 3D warehouses, and can be merged into the architectural design file. The final 3D data can be printed three-dimensionally by any AM process.

Among others, additive manufacturing processes can provide three-dimensional models of anyone's house, e.g. your hometown, friend's sites, prominent buildings, bridges, and more. By means of software, individual coloring is possible to emphasize individual attributes (Figure 4.36).

Figure 4.36 Urban landscape: landmarks emphasized by individual coloring
(Source: 3D Systems)

Figure 4.37 shows a model of the mountainous region "Berner Oberland" (Bernese Highlands), Switzerland. It is part of a 5 m × 5 m large relief that was displayed on the GEO Summit (2016) in the city of Bern, Switzerland. The 3D relief was printed from PLA by means of a personal printer and assembled at the "Institute for Further and Media Education". The numerous white relief-like prints produced were used as a 3D projection screen. With the help of a beamer, different textures or pictures were displayed: static views obtained from Google Earth or animations from the actual weather service.

Figure 4.37 3D relief: Projection onto a 3D printed white 3D screen. Screen produced by a personal printer, PLA
(Source: Institute for Further and Media Education Bern)

The Aachen Cathedral (Kaiserdom) – also called the "Aachener Münster" – in the city of Aachen, Germany is one of the most important sacred buildings north of the European Alps. Built in the last decade of the eighth century, for more than 200 years it was the building with the highest arches and the widest span of its vaults. The central section of the cathedral is the Carolingian octagon, where most of the German kings and emperors were crowned.

The "Aachener Münster" was included in the list of the UNESCO world heritage sites in 1978 as its first German monument. The 35th anniversary in 2013 was the cause for a project of the high school "Einhard-Gymnasium" and the "Aachen University of Applied Sciences". The task was to document the 1200 years of building history. With the help of photos and ground plans from the internet, the proportions were fixed and after an extra Autodesk Inventor course were transformed into a 3D CAD model. From this, first a 10-inch full model was printed. In addition, the most prominent attached chapels were manufactured separately in color and linked to the building of origin. The resulting "3D puzzle" provides the basis for the study of the sequence of modifications and extensions of the cathedral, carried out over centuries.

For the additive manufacturing, the extrusion process (FDM) was chosen (see Section 2.1.3 "Extrusion/Fused Layer Modeling").

The apparent disadvantage of the extrusion process, the clearly visible "sausage-like" texture, was deliberately used to mimic the structure, mainly of the roofs. The thereby achieved effects resulting from the additive manufacturing process would not have been feasible with justifiable expenditure directly by CAD. The black material emphasizes the value of the model. Figure 4.38 shows the total complex (left) and a detailed section of the roof (right).

Figure 4.38 Model of the Aachen cathedral, extrusion process (FDM); CAD data reconstructed from photos and revised for model building; scale approx. 1:500; the different colors mark different epochs (left); detailed section of the roof with deliberately used extrusion structure to display the texture (right) (Source: GoetheLab, University of Applied Sciences, Aachen)

■ 4.9 Miscellaneous Applications

As the only prerequisite for the production of parts by additive manufacturing is a 3D data set, a 3D object can be made from any suitable data source. Therefore, nearly every branch of industry is using AM. To highlight this and to motivate new applications, some examples of special applications are described as follows.

4.9.1 Mathematical Functions

Even mathematical functions can become attractive when they are displayed as three-dimensional objects. Figure 4.39 (left) shows the CAD rendering of so-called "minimal surfaces" and Figure 4.39 (right) displays the physical 3D model of this function, made by additive manufacturing, in particular by stereolithography. All additive processes that provide good surface quality and preferably do not need supports can also be used for this purpose.

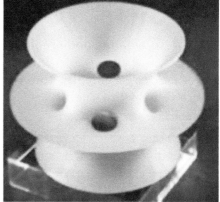

Figure 4.39 3D model of a "minimal surface": CAD rendering (left); stereolithography model (right)
(Source: David Hoffmann and Steward Dickson)

4.9.2 3D Decoration Objects and Ornaments

With the powder-binder process (3D Systems/Z-Corporation) – also called Color Jet Printing (CJP) – an additive manufacturing process is available that not only allows the display of nearly any geometrical form, but also provides continuous coloring. In this way decorative colored objects or ornaments like the "serpent knot" (Figure

4.40) are made. In printing such textures, the limits of the STL format quickly become evident. Bitmaps, transmitted by ply, VRML, or at the best by AMF format (see Section 3.1.2 "Data Structure, Error, and Repair") should be used.

Figure 4.40 Decorative object "serpent knot", colored 3D printing (powder-binder process, CJP) (Source: 3D Systems)

4.9.3 Aerodynamic and Freeform Objects

One of the biggest advantages of AM is the possibility to generate any freeform surface as a physical object. Some examples are shown in Figure 4.11 (right) and Figure 4.21.

Therefore, AM is successfully applied in aerodynamics. Today the conception of a new aircraft is carried out by means of 3D CAD and simulation programs, starting from the first sketch. Apart from computer graphics and renderings, which provide a good impression of the basic idea, a scaled model is used for better understanding of the concept and to improve the feeling for the proportions. Figure 4.41 shows the concept of a single engine aircraft as a rendering (left) and as the corresponding AM model (right). The model was made by means of laser sintering and shows the freeform surfaces without influence of post-processing.

Figure 4.42 shows a set of accessories for the improvement of the aerodynamics of a racing car. Like the racing car itself, they were reduced in scale for testing in a wind tunnel. The test provides a quick and inexpensive first impression and helps to focus on the essential consequences, thus reducing the test period and the costs. For manufacture of the parts, stereolithography was chosen with regards to the moderate mechanical load.

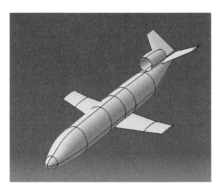 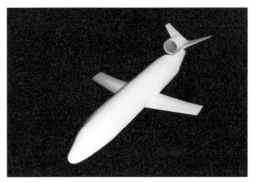

Figure 4.41 Concept of a single engine aircraft: CAD rendering (left); AM model, laser sintering, polyamide (right)
(Source: Philipp Gebhardt, University of Stuttgart)

Figure 4.42 Aerodynamically optimized accessories with freeform shapes for a racing car: installation in a wind tunnel for testing; parts made by stereolithography
(Source: 3D Systems)

■ 4.10 Conclusions

As each industrial or craft business already uses or is about to use computers to generate or to improve their products by means of 3D CAD, the most important precondition to apply additive manufacturing processes (the entry into digitalization) is already fulfilled or at least does not form a big obstacle anymore.

As a consequence, currently almost everybody is a potential user of additive manufacturing.

The presented examples show clearly that there is no fixed linkage between additive manufacturing processes and branches of industry. The decision regarding which AM process to use has to be taken from case to case. Sometimes this gets difficult, because in some cases satisfactory results can be achieved by different AM processes, while in others only one process is applicable. Therefore, the suitable additive process has to be chosen depending on the requirements of the parts, and not depending on the branch of industry.

■ 4.11 Questions

1. **Why can AM prototypes and AM products be manufactured by the same machines, with the same processes, and from the same material?**

   Whether a part is a prototype or a product does not depend on the used machine or the processed material, but on design and construction. The part is a product if all design requirements are met. This surely is the case when a design suitable for AM, based on the AM material data, has been carried out.

   If a part is designed based on a classical series production (e.g., for manufacture by means of injection molding) and the corresponding series material is applied, but is manufactured by means of AM, and AM material is applied, it is a prototype independent from the used AM machine and AM material. This prototype may be more or less in accordance with the design, but in any case it remains a prototype.

2. **What kind of additive manufacturing process is preferably applied for the cores of sand-casting? Why?**

   Polymer-bound form-sands are preferably applied by means of laser sintering, non-bound sands by means of 3D printing, and polymer-binder for manufacturing of molds and cores for sand-casting.

   The materials are to a considerable extent in accordance with those applied for traditional sand-casting.

3. **Which branch of industry mainly uses stereolithography?**

   The application of a distinctive additive manufacturing process, e.g. stereolithography, does not depend on the branch of industry but on the characteristics of the parts to be manufactured. If fine, delicate details need to be incorporated, a good surface quality has to be achieved, and the part will be exposed only to low temperatures, stereolithography is the preferred AM process. This is valid for any branch of industry.

4. **Which materials can be processed by metal laser sintering or metal laser melting?**

Laser melting produces dense metal parts. The processes are designed to process commercially available metal powders. A wide variety of mixtures, resulting in alloys, is available and can be used for manufacturing, as long as the powders are qualified, or the user carries out the qualifying by himself/herself. The powders offered by the manufactures of AM machines can be considered as proven. As materials steel, stainless steel, tool steel, CoCr-alloys, titanium, aluminum, and others are available. Magnesium and copper are about to enter the market.

5. **Which AM processes can be applied to manufacture colored parts?**

If only one color is required, the AM material of most plastic processes can be dyed after printing. For extrusion processes (FDM), colored materials are available, and colors can be changed quickly. Only one color can be printed at once. But as multiple colors are available, a multi-color part can be assembled after printing.

If stereolithography or sintering is required, the complete build chamber has to be filled with the desired material.

Multiple colors, even with gradations, can be processed simultaneously by polymer printing, applying the PolyJet process (Stratasys/Objet). In this case, additionally multiple materials can be processed simultaneously.

Multi-color parts can also be manufactured by means of the 3D (powder-binder) printing process (Color Jet Printer, CJP, of 3D Systems). It allows the printing of bitmap-based textures. The depth of color penetration, however, is not very high and only one material can be used.

6. **For many years architects have been using scaled models and have developed considerable model manufacturing skills without applying AM. Which improvements in architectural model manufacturing can be achieved by using AM?**

All parts of an architectural model that can be made by conventional technologies such as cutting and milling should further on be manufactured in this way. AM should be applied preferably if freeform surfaces or very detailed parts are required. Often, models are assembled from parts that are manufactured by AM as well as in the traditional way. If colored objects are required and manual painting should be avoided, the use of "full-color" 3D printing is recommended.

7. **AM processes are often only partly suitable for the production of small series or parts with defined properties as transparency or elasticity. Which application options are offered by AM in such cases?**

   AM provides a very good geometrical model within a brief time. If the required properties cannot be met directly by this part, casting or secondary processes are used, such as for example vacuum casting.

8. **How do toys for children and adults differ? What does it mean with regards to AM?**

   Toys for adults primarily are made for presentation purposes and require even in the prototype version extremely fine details and delicate structures. Therefore, polymerization often is used to meet these requirements. Toys for children have to be more robust, which often has to be proven already in the prototype phase. In this case, mostly sintering or extrusion of plastics is applied.

9. **What are the two approaches to manufacture medical implants for skull operations (cranioplastics)?**

   Both approaches are based on the 3D reconstruction of the CT data set of the skull including the defect. Either the master model for precision casting is generated by means of polymerization from wax-like plastics, or by sintering from amorphous plastics. Afterwards the manufacture of the titanium implant is carried out by precision casting.

   Alternatively, the implant can be sintered directly by laser melting (SLM) from titanium. The direct AM manufacturing is faster, whereas the lost wax process cheaper. If larger modifications should be necessary, the wax-master model can be modified quickly at low cost and casting can be repeated.

10. **What correlation exists between AM and conformal cooling?**

    Conformal cooling requires three-dimensional cooling channels that follow the surface of the tool cavity as closely as possible. Channels like these cannot be drilled or inserted before casting – apart from a few exceptions. Layer oriented additive manufacturing processes are the only method to manufacture these channels. Therefore, metal laser melting processes (SLS) are used.

References

[1] Gebhardt, A.; Hötter, J.-St.: *Additive Manufacturing: 3D Printing for Prototyping and Manufacturing,* Carl Hanser Verlag, Munich, 2016

5 Perspectives and Strategies of Additive Manufacturing

This chapter deals with new strategies for engineering design, development, and production of products that can only be made, or can be produced more simply, by means of additive manufacturing. In addition to Chapter 4 "Applications of Additive Manufacturing", Chapter 5 can be considered as a systematic approach to the application of additive manufacturing processes.

Based on the potential of additive manufacturing processes, strategies for the manufacture of new products with alternative or new design elements are demonstrated. The aim is to motivate the reader, on the basis of examples, to search for applications in their own field, and thus to support the application of additive manufacturing processes and finally to develop and manufacture better products in a shorter time.

This different and new way for the development and manufacture of products moreover offers the chance to build up a completely new customer–manufacturer relationship and consequently to discuss and establish new developing strategies.

This chapter builds on the knowledge of the five fundamental process families and the process chains of additive manufacturing according to Chapter 2 "Additive Manufacturing Processes/3D Printing".

■ 5.1 Potential of Additive Manufacturing

As described in Chapter 1 "Basics of 3D Printing Technology", 3D CAD data sets are directly transferred by means of AM into real (or physical) 3D objects[1] that can be used as prototypes or products. Prototypes are three-dimensional objects for visualization, or sculpture-like objects, but may also show selected functions. However, products (target or series parts) have to show all functions that have been defined during the design phase.

---

[1] Under the basic conditions according to Chapter 1 "Basics of 3D Printing Technology".

The parts can be "positives" or "negatives" (tools or models). As already discussed in Chapter 4 "Applications of Additive Manufacturing", the range of possible applications with regard to additive manufacturing is large. The examples mentioned in this chapter are assigned to selected branches of industry, to make orientation easier for potential users.

The as-yet unexploited possibilities of additive manufacturing offer the potential of another industrial revolution[2]. Additive manufacturing allows anybody to generate parts or products showing all imaginable geometries, to manufacture them from any desired material and in any quantity, and to do this in any place, and – if required – even simultaneously in different places. Thus, the start of an epoch of customized mass production, capable of revolutionizing today's industrial production and society, is identified.

The potential of additive manufacturing to initiate an "industrial revolution" is based on three of its fundamental properties:

- Firstly, additive manufacturing processes deliver very complex geometries that cannot be manufactured by most of the traditional manufacturing processes.

- Secondly, already today occasionally and widespread in the near future, materials can be varied to a large extent during the manufacturing process, allowing the generation of totally different part properties and corresponding product properties (graded materials).

- Thirdly, the direct digital manufacturing process allows simultaneous production of different products in one manufacturing sequence without geometry-related or product-dependent tools, wherever the production machine is located.

It is not necessary to emphasize further that the success of AM is mainly based on the fast development of systems for 3D data generation and handling, including 3D CAD systems, scanners, internet-based 3D data sources, and 3D data libraries.

Below, the essential AM strategies are discussed and illustrated by examples. There is no claim that the examples are exhaustive, or that the order of their appearance is an assessment of their significance. The aim is to prompt the reader to look for similar application potential in his or her field of experience.

---

[2] Whereas the first "industrial revolution" at the beginning of the 18th century is defined as transition from the agricultural to the industrial society, further important changes are judged differently. The introduction of electricity and mass production at the beginning of the 20th century is called the second industrial revolution (G. Friedmann), while the computer-based additive manufacturing processes often are considered as a third industrial revolution. The aspiring microelectronic industry has a comparable potential. The definitions are not final and not standardized.

5.1.1 Complex Geometries

The adjective "complex" is applied to parts that can be manufactured directly as complete pieces by the application of additive manufacturing processes, whereas their production by means of conventional processes require multi-stage production processes, complicated tools, and final assembly.

As discussed before in Section 4.7 "Medical Engineering" and shown in Figures 4.28 and 4.29, a human skull, geometrically one of the most complex objects imaginable, can be manufactured directly and as a complete part by means of additive manufacturing. This could not be achieved by conventional manufacturing processes or would require unacceptable expenditures with regards to time and money. The figures clarify that depending on the requirements with regards to the part, different additive manufacturing processes can be applied.

Comparable complex geometries include tools or tool inserts with conformal cooling [1]. The idea to provide conformal cooling channels is not really new. With conventional methods, using drilled, straight channels with circular cross sections, which are not able to follow the contour of a cavity close to its surface, the aim is hardly realizable. Additive manufacturing enables design and manufacture of customized cooling channels, and even three-dimensional cooling (channel) grids, which can improve the productivity (cycle time reduction) of the tool significantly. In Figure 4.27 a comparison between conformal cooling channels made by additive manufacturing and drilled, straight cooling channels is depicted.

A complex cooling grid is shown in Figure 4.27 as a CAD rendering, as well as a cut-away AM part. The mold insert was made by means of selective laser melting from steel powder.

Another example for a very complex geometry, again from the medical field, is the model of bronchial tubes shown in Figure 5.1. The model was used as lost core for the manufacture of a transparent test duct for flow tests in human respiratory organs [2]. The data set can only be generated from living individuals (otherwise the tubes collapse) by means of a CT scan with subsequent 3D reconstruction. The core was manufactured by the powder-binder process (3D printing) and then cast in silicone.

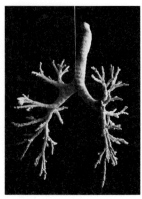

Figure 5.1 Model of human bronchial tubes in scale 1:1: 3D printing (powder-binder process); model after AM (left); object after removal from the machine and post-processing (right)
(Source: GoetheLab, University of Applied Sciences, Aachen)

To achieve clean ducts, the core had to be removed completely after the solidification of the silicone. Therefore, it was made with the minimum possible amount of binder. The result was a very fragile part that required very sensitive handling and cleaning.

Another example for the nearly unlimited range of geometrical elements that can be manufactured by additive manufacturing comprises the 3D net or grid structures shown in Figure 5.2. The photograph proves that the manufacture by AM is able to produce fine, delicate, and complex elements not only from plastics (see for example Figure 5.13), but also from metal.

Figure 5.2 3D net structures: selective laser melting, CoCr alloy; parts after removal from the machine and slight grid blasting
(Source: GoetheLab, University of Applied Sciences, Aachen)

The net structures in Figure 5.2 are not products, but demonstrate the possibility to manufacture implants, elements for heat transfer, stiff but lightweight structures, and other complex geometries by additive manufacturing. The complexity of grid structures can be varied nearly without limits. Additive manufacturing processes allow – based on the optimization of the topology – the generation of grid structures within one part that show different spacing, as well as varying forms of structures.

Figure 5.3 shows additively manufactured spring elements from stainless steel with integrated net structure as demonstration objects. For demonstration purposes, the wall of the hollow construction of the springs was partly left unprinted. The springs are not products but demonstrate the possibilities of additive manufacturing.

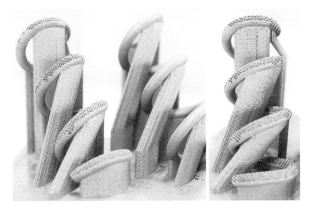

Figure 5.3 Spring elements with integrated net structure: selective laser melting, stainless steel; parts after removal from the machine with supports
(Source: IwF – Institut für werkzeuglose Fertigung (Institute for toolless Fabrication)/GoetheLab, University of Applied Sciences, Aachen)

5.1.2 Integrated Geometry

For today's non-additive manufacture, often the complex geometry of a product has to be simplified. Usually such a part is split into single elements that can be manufactured in accordance with the restrictions of the chosen manufacturing process. In these cases, a final assembly is required to complete the product.

A very big advantage of additive manufacturing is that it allows the generation of highly complex geometries directly. This means from the strategic point of view that a product is assembled virtually with the aim to reduce the number of tools significantly, to eliminate production bottlenecks by parallel manufacture, to eliminate final assembly, and to reduce storage capacity and management.

The blood container of a blood centrifuge (Figure 5.4, left) for the preparation of blood tests is a very good example. In Figure 5.4 (center) it can be seen that the container, made by means of tooled (traditional) manufacture consists of three single elements (1, 2, 3). For the manufacture by plastic injection molding, accordingly three tools are required. Additionally, the elements have to be assembled. While the centrifuge itself is standardized, the containers usually belong to the hospital and vary according to hospital policy and local or governmental regulations. Unfortunately, in many cases the number of the required containers is too small to justify the costs of the tools. Thus, small companies are excluded from sections of the market. Applying additive manufacturing, the single elements (1 to 3) can be integrated into one geometry and then manufactured as one AM part (Figure 5.4, right).

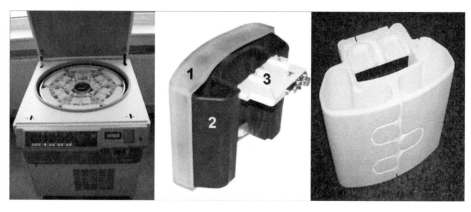

Figure 5.4 Blood centrifuge: overall view (left); blood container, made from three plastic parts (1, 2, 3) by injection molding (center); blood container generated by means of AM as one part by laser sintering from PA (right) (Source: Hettich/EOS GmbH)

As no tools are required, small series and even one-of-a-kind parts can be produced economically by means of an AM process (sintering of polyamide). The manufacturing costs per container are higher when using additive manufacturing processes instead of injection molding, but are far beyond the break-even point for tool-bound production with regards to the same number of produced containers.

Traditionally complex cores for casting are assembled from several single elements. This manual procedure is very expensive and time-consuming. Moreover, there is a risk of alignment errors occurring during assembly of the elements that result in staggered elements and consequently cause defects in the casted parts.

Further examples for complex parts are displayed in Figure 5.5. The vortex element for a jet turbine (Figure 5.5, left) was manufactured by laser melting (SLM) from hot work tool steel. It is a relatively big part with the dimensions 298 × 120 mm. After post-processing it is ready for installation.

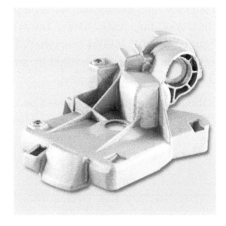

Figure 5.5 Vortex element (left) for the combustion chamber of a jet turbine: SLM hot work
tool steel. Base plate (right): aluminum, dimensions 30 × 100 × 50 mm
(Sources: Concept Laser/GE (left); EOS GmbH/Morris Technologies (right))

The base plate of a mirror (Figure 5.5, right) was made from aluminum (AlSi12) by
means of SLM in one piece with the dimensions 30 × 100 × 50 mm. The traditional
manufacturing method for both parts is a sequence of casting, welding, heat treatment, and mechanical processing.

The principle of "integrated geometry" can be applied with regards to all processes,
all kinds of parts, and all materials in all industrial branches.

5.1.3 Integrated Functions

Applying traditional (non-AM) manufacturing, the final product mostly needs to be
assembled from semi-finished parts according to the desired functionality. AM
allows the directly manufacture of complex parts so the assembly procedure can
consequently be skipped. This includes "integrated functions" by using the flexibility of the material.

Two functional principles are widely applied: living hinges (film hinges) and snap-
fits. Both use the elasticity of materials and belong to the class of solid joints (also
called flexures), as well as articulated arms (of the hinge type). For this, mostly a
rigid part including all elements is manufactured additively, in the form of one
piece. It provides the desired movability by removing internal supports after the
build.

These functions have to be defined in the design phase by skillful positioning of
the single elements on the build platform. While living hinges or snap-fits can also
be manufactured by plastic injection molding, articulating elements can hardly be
made by tool-bound processes.

As an example, Figure 5.6 shows a planetary gear train with helical and double-helical toothing, respectively. The whole gear was made in one process by SLM (Figure 5.6, left) or FDM (personal printer) (Figure 5.6, right). The position of the toothing elements was verified by small connecting layers that were removed by operating the part (manually).

As the part cannot be assembled because of the helical toothing, it is not possible to manufacture these elements conventionally.

Figure 5.6 Functional elements, made in one piece (planetary gear train, spring): SLM, metal (left); FDM, plastics (right)
(Source: GoetheLab, University of Applied Sciences, Aachen)

The adjustable air-outlet grille of a passenger car in Figure 1.9 was made in one build process by stereolithography. The necessary gap for the mobility of the joints was generated by leaving one or two layers uncured and thus not solidified (see Chapter 6 "Materials and Design"). In the course of the cleaning works, the material in these areas was removed by solvents and by moving of the elements, thus releasing the respective gaps. Afterwards the joints were ready for use.

This method, to manufacture joined, but movable elements, can be considered as AM construction principle. It is preferably used in plastics sintering processes, but may also be used for polymer printing, 3D printing and in extrusion processes (FDM). It also can be applied for metal printing.

If processes require supports, support material, which will be removed after the build process, is applied to form the gaps needed for the later movability of the joints. Water-soluble supports are of advantage. The adjustable wrenches made from plastics shown in Figure 5.7 comprise good examples for movable elements such as guides, gears, and, in this case, worm gears. They also demonstrate that different AM processes – in this case polymer printing (above) and extrusion (FDM, below) – can be applied.

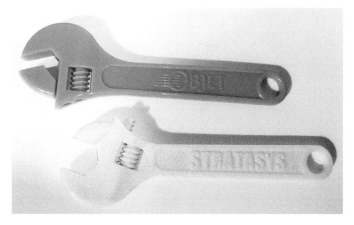

Figure 5.7 Adjustable wrenches, demonstration models: polymer printing, acrylate (above); FDM extrusion, ABS (below)
(Source: Objet (above); Dimension/Stratasys (below))

The 3D-printed Sarrus mechanism in Figure 5.8, which transforms a circular movement into a linear one, is another example of additive manufacturing of products consisting of various elements with integrated geometrical functions, in one piece. The linkages, including all elements – with the exception of the last hinge – were printed in one piece.

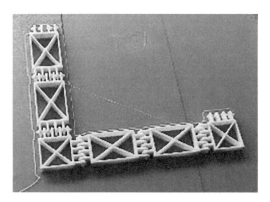
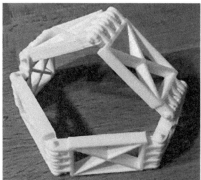

Figure 5.8 Sarrus mechanism: printed in one piece, FDM; after printing on the printing bed (left); clipped together (right)
(Source: Jon Hollander)

The engaging cable clip in Figure 5.9 was additively manufactured in one piece.

A comparable traditional cable clip consists of different plastic and metal parts, bolts, and rubber inserts, in order to fix and protect the plastic cable [3]. A cable clip suitable for AM was designed as a single plastic (AM) part with two movable frame clasps, one for the supporting tube and a second for the cables.

Figure 5.9 Foldable cable clip: made in one piece by sintering from polyamide; closed state (left), open state (right)
(Source: EOS GmbH)

Both frames can be closed and opened by snap-fits. Even the traditional elastic pads, made from rubber, are integrated into the clip. To replace the rubber pads, the stiff plastic material was formed like saw teeth, to firstly cover the existing gap and secondly provide the required elasticity for fixing of the cables. The basic version, shown in Figure 5.9, can be varied with regards to the basic shape, the diameter of cables, the number of cable cords, etc. Different concepts can be generated in one build process. After the build process and finishing by slight grit blasting, the parts are ready for use.

Figure 5.10 Fin-type grippers: distortion depending on the torque applied to the base plate; filled polyamide
(Source: Festo)

Another variation of this principle is the fin-type grippers of company Festo, which represent a different type of construction element, the solid joint or flexure. It comprises a stiff, triangle-shaped element (Figure 5.10), which can be distorted specifically by applying a torque at the clamping point. Integrated spacers, fixed according to the hinge method (Figure 5.9), cause the distortion that is dependent on the direction of the torque (Figure 5.10). This allows a grasping movement and thus manipulation of the object to be handled with minimum normal force.

Figure 5.11 shows the grippers as elements of a bionic handling system, which resembles an elephant's trunk.

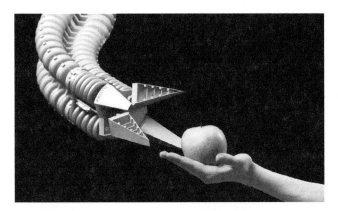

Figure 5.11 Bionic handling system (elephant's trunk): consisting of sintered elements according to Figure 5.10; laser sintering, aluminum-filled polyamide (Source: Festo)

The Festo Company specializes in pneumatic systems. Therefore, corrugated hose-like, sintered elements were used for the elephant's trunk so it can be operated by compressed air and thus be directed. Additive manufacturing allows variation of the wall thicknesses according to the required distortion.

The complete bionic handling system was presented at the Hannover Fair in 2011. It integrates the grippers and the pneumatic elements and is controlled by external Bowden cables and compressed air (Figure 5.11).

The manufacture of flexible products like textiles and fabrics, for example handbags or clothes, by means of additive manufacturing was first presented by FOC (Freedom of Creation) some years ago. The use of hard materials for the manufacture of smooth products was interesting, but the design resembled more a chain-mail rather than modern clothes. In 2011 a few young women of "Continuum Fashion" revived this idea and presented the world's first AM bikini (Figure 5.12). The structure consists of small rings interlinked by thin ribbons as flexible connections. The fabric is manufactured in one build process and in one piece by laser sintering. The aesthetically attractive design also meets structural requirements such as durability, flexibility, and, above all, fit. Continuum Fashion calls it a completely new material. Due to the rigid sintered polyamide elements the product is designed to be worn at the beach only temporarily. The hydrophilic behavior of the material does not restrict its use, as the slight geometrical deviations due to water absorption affect neither the fit nor the elasticity. The model has to be protected from UV radiation by a special coating.

The project is a further step toward additive manufacturing of clothes.

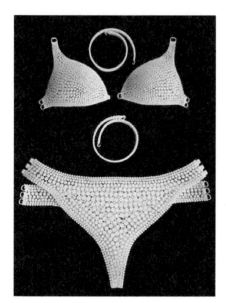

Figure 5.12 AM for clothes: first 3D-printed bikini worldwide; laser sintering, polyamide (Source: Continuum Fashion/Shapeways)

Figure 5.13 Red dress: laser sintering, polyamide; total view (left); close-up (right) (Source: fastcodesign.com)

At the design exhibition #techstyle (July 2016) at the Museum of Fine Arts, Boston, MA, impressive pieces of clothing made by a 3D printer were presented. Among others, Nervous System, a design studio, based in Somerville/Boston, MA, presented a new kinematic, 3D-printed dress (see Figure 5.13).

The dress, inspired by natural elements like petals, feathers, and scales, leads into new dimensions in the field of customized products through its "flowing kinematic".

5.1.4 Multi-Material Parts and Graded Materials

The majority of parts discussed and shown in this book are made from one (ideally) isotropic material. Isotropic material behavior is considered as mandatory, because nearly all non-AM-based manufacturing processes like mechanical processing or casting generate (nearly) isotropic parts. Accordingly, nearly all construction rules are based on isotropic material behavior.

Additive manufacturing, however, has the potential to abolish this restriction. It allows the production of parts made from several materials, as well as their continuous changing within the part.

The discussion starts with actual applications, but also includes new processes that will enter into the market in the near future.

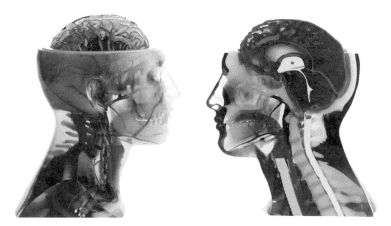

Figure 5.14 Polymer printing: multi-material and multi-color parts, made by the machine
Stratasys J750
(Source: Stratasys / Objet)

The polymer printing process, which is already established on the market (PolyJet, Stratasys/Objet; see Section 2.1.1 "Polymerization", Figure 2.6) is able to process simultaneously two materials with a defined ratio of one to the other, and to simulate thereby a two-components injection molded plastic part. As an example, in Figure 2.7 (right) a wheel is shown that consists of a hard rim and an elastic tire that was made by polymer printing in one piece. In a further development, the machine "Stratasys J750" is capable of processing simultaneously up to 360,000 colors and up to six materials and thus manufacture multi-material and multi-color parts.

The development of machines for multi-color parts is also extended to 3D printers that work according to the extrusion process. Today, FDM printers with two extrusion heads are simultaneously processing build and support material. In the field of personal 3D printers these are, for example, the "Ultimaker 3" or the "CubePro" of 3D Systems.

Figures 1.7 and 4.17 show parts that were colored individually during the AM process, and whose part property "color" was varied continuously. In principle, instead of different colors other material properties could also be varied. Thus, a process may be conceivable that unites a variety of different properties in one part.

The powder-binder process and all other AM processes that put together the build material voxel by voxel to form a part are fundamentally suitable for realization of each voxel with different properties. In the near future it will be possible to design and manufacture parts according to properties as tensile strength, elasticity (Figure 2.7, right), transparency, electrical and thermal conductivity, and others, whereby it is possible to vary the properties within the cross-section, and to adapt the material to the design rules in this way.

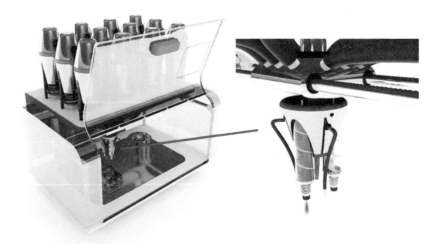

Figure 5.15 Additive manufacturing of food (food processing): 3D print concept for food, "The Cornucopia", MIT; printer set-up (left); details of print head (right) (Source: Diane Pham/Inabitat.com)

An evolving application of additive manufacturing processes that deals with multi-material processes of organic materials is the production of food, called *food processing* or *food printing*. The first attempts go back to the cold "Fab-at-Home" extrusion processes, which deposited prefabricated pasty food by means of electrically controlled syringes. The extruded material became stable after processing and resulted in specifically shaped, edible objects, e.g. from beaten egg white and sugar. At the University of Applied Sciences Aachen, Germany, processes have been developed that work with chocolate or "gummi bear" mass. These are still under development.

In 2013, the Massachusetts Institute of Technology promoted the 3D print concept "The Cornucopia", claiming to be a personal food factory (Figure 5.15).

The food printer uses containers for food ingredients. Clients place their order on the spot and observe the selection, the mixing, and the preparation of their food. The composition of the different ingredients is carried out by extruder. For this purpose, the extruders are equipped with heating and cooling pipes. Food printers process very exactly with regards to reproducibility of the food's quantity and its composition. They also take care of the precise dosing of the food components like fat, lactose, calories, carbohydrates, and others. Whether in reality waste is reduced has to be evaluated in future investigations. The preparation of the ingredients and cleaning of the machine have to be considered. Figure 5.16 shows the food printing process by a fabber.

Figure 5.16 Food printing in process: modified fabber, extruding food
(Source: 3dprintingindustry.com, 3DigitalCooks)

Further multi-material machines and processes such as the "Maskless Mesoscale Manufacturing process M3D™" of Optomec or the "3D Bioplotter ™" of Envision-TEC are already capable of producing electronic circuits, drugs, or human bones and tissue. At the beginning of 2010 scientists of "Organovo" generated the first printed veins by means of the "MMX Bioprinter" of NovoGen. Even printing of organs is emerging as an option on the horizon.

■ 5.2 Strategies of Additive Manufacturing Processes

Lately, clients are increasingly asking for individually manufactured unique products. The customers reject mass products more and more, and demand unique and signature products, tailor-made to their needs. This is called *customization, customized production*, or *mass customization*.

It is very difficult to meet these requirements under the paradigm of mass production. As the commercial success of a production is closely linked to the number of manufactured products per unit time, high numbers of items are demanded and realized by corresponding tools (molds). However, by means of molds identical items are produced.

Modifications of the product or even smaller changes are avoided as far as possible, due to costs and the necessary time for the modification of the tools or tool inserts, and the re-adjustment of the production machines.

Additive manufacturing has the potential to effect a real change, because tools are not required, and thus the requirements of an individual production are met perfectly. By means of additive manufacturing, different parts can be generated in one build process. Each part can be either a one-of-its-kind part or it may belong to a small, medium, or even large series.

AM leads from mass production of identical parts to mass production of different or individual parts. For the customization by means of additive manufacturing two ways are open. It can be realized under responsibility and supervision of the manufacturer, or the client can be involved directly or indirectly. The customization and the production may also be completely in the hands of the client – this is the revolutionary aspect of additive manufacturing that may be considered as aparadigm change in the product genesis (product development and manufacturing).

Although, from the point of view of manufacturing, customized design is independent from customized production, customization always involves the client as designer, at least to a certain extent. Each serious discussion has to consider both design and manufacturing. If the responsibility is with the manufacturer, and the production is carried out in his/her workshop, the procedure is called *customized mass production*. If customers carry out the production by themselves and within their own facilities or with private responsibility, the procedure is called *personal fabrication*. The design (layout) can be carried out with the responsibility of the manufacturer or the customer or in an intermediate state.

If the personal fabrication is carried out by a large number of private customers, this is a good basis for internet-based production networks, located in a cloud.

Thus, totally new product development and production strategies are emerging, named *distributed customized production* or *coproducing*.

Independent from the preferred strategy, AM machines, as well as any other kind of production machines, require investment, and therefore have to earn a return on investment. To achieve this, the machines have to produce as many parts per unit time as possible. Whether the process is economical or not is determined to a rather small extent by the engineering or the design of the parts. The organization of the AM-based customized mass production is decisive.

5.2.1 Customized Mass Production

Customization means to adapt a product to the requirements of a special customer or a special group of customers. Customization can be carried out in respect of quantity or quality, which means either to manufacture one-of-a-kind parts or small batches, or to modify appearance, geometry, or function of a part. The design of a part can be modified depending on the taste of a group of individuals, which is called *individualization*. If the part is manufactured to meet the requirements of a special client, this is called *personalization*. If the customer influences the design creatively, this is called *active personalization*.

The process of individualization is closely linked to the (engineering-) design procedure, while additive manufacturing is a manufacturing technology. As discussed in Section 3.1 "Data Processing and Process Chains", AM represents an integrated design and manufacturing process. Therefore, its main influences cannot be considered separately.

5.2.1.1 One-of-a-Kind and Small Batch Production

One-of-a-kind production, or the production of small batches, represents the quantitative aspect of individualization, i.e. the realization of the individualized production by the number of items. The effect of individualization is characterized by production on demand, while the product remains unchanged. Additive manufacturing allows the production of an arbitrary number of items with respect to the customer's needs: just one item, one item per time unit, or a defined number of items several times a year. To realize such a production based on tools requires operation of the process repeatedly, even for uneconomical small batches, or to produce the whole series at once. This would result in an economical number of items that however would be far above needs. The overproduction would have to be stored (and financed).

An example is the leak detector in Figure 5.17. The customer orders a small batch of equal products every couple of months. The production is carried out by AM, because the quantity of items is too small to justify the manufacture of a tool and the subsequent plastic injection molding process. As a solution, the parts are manufactured by laser sintering, get a final black coating, and are equipped with the electronics, a display, sensors, and the handle.

Figure 5.17 Leak detector: laser sintering, final coating, and assembly
(Source: CP-GmbH)

5.2.1.2 Individualization

Individualization and personalization (see Section 5.2.1.3 "Personalization") are approaches to "customized (quality) products". By means of additive manufacturing a larger number of variations of a product can be realized in order to meet the requirements of different groups of customers based on one initial construction. If the product is designed by the manufacturer, this strategy is called *individualization*. As example, in Figure 5.18 and Figure 5.19 various AM master models for the production of jewelry are shown. They present two different AM-based solutions to manufacture the final part: direct wax-printing of a scaled master model for investment casting (Figure 5.18) and direct manufacture of the jewelry by selective laser melting of precious metals (Figure 5.19).

In both cases, the customers have not been involved in the design personally. The products were designed by professional designers based on suitable marketing methods.

Both product variations were made on one build platform and in the same build process.

The direct manufacture of jewelry can also be carried out by means of another additive manufacturing process whose fundamental principle is based on melting or sintering of metals.

Figure 5.18 Customized production; individualization: master models for investment casting, made by direct wax-printing (Source: EnvisionTEC)

Figure 5.19 Customized production; individualization: direct manufacture of jewelry by selective laser melting of gold (Source: ReaLizer GmbH)

Each AM process requires, as for any traditional (non-AM) manufacturing process, finishing. This is often considered as a disadvantage and is the reason for the development of automated finishing procedures. Such systems are under development in the design phase and will have big advantages for industrial applications.

Seen from the craftsmen's point of view, the required finishing can also be considered as an advantage. Additive manufacturing does not "eliminate" traditional workplaces but provides perfect semi-finished products (or green parts), which are formed, by the skills and knowledge of the specialists, into perfect (and unique) jewelry. Thus, additive manufacturing supports the creativity, competitiveness, and further development of the craftsmanship, as well as small and medium-sized enterprises (SMEs).

Individualized products are to be found in nearly every branch of industry. Examples have already been shown in Chapter 4 "Applications of Additive Manufacturing".

The range of application of additive manufacturing processes is versatile and expands rapidly to new branches, like the furniture industry. An impressive example is the chair carried out by the French designer Patric Jouin shown in Figure 5.20 (above). The structure visualizes the flow of forces. The chair could not be manufactured by traditional processes.

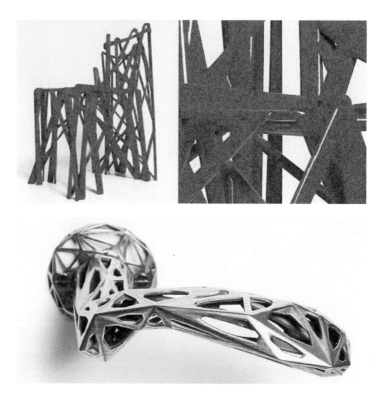

Figure 5.20 Customized production; individualization: furniture; chair from the "solid series" of Patric Jouin (above). Individualized door handle, made from stainless steel (below)
(Sources: R. Guidot [4] (above); i.materialise (below))

The company i.materialise offers a kit that allows designers and 3D modelers to manufacture their own individualized door handles and to print them from stainless steel; see Figure 5.20 (below).

Individualized products are not inevitably one-of-a-kind products and in principle can be manufactured by traditional manufacturing processes or by means of AM processes. If, due to more intensive individualization and an increasing number of variations, the production volume of one variation should decrease, the break-even point of the production costs moves to very small numbers of items. Then, only by means of additive manufacturing can an economical production be achieved.

5.2.1.3 Personalization

Personalization is characterized by the development and manufacture of a unique or only-once-manufactured product (one-of-a-kind product; one-offs). Its details are determined by a certain person, mostly directly by the customer. The customer determines or influences to a considerable extent the product design either by his/her biometrical parameters (passive personalization) or by his/her own creative potential (active personalization).

In general, personalization requires a basic design engineering or a basic conceptual design and a manufacturing chain that determines the product and the potential for personalization. The final personalized product results from the adjustment of this chain and the software behind it according to the needs of the respective customer. The total process design, as well as production, will be finally carried out under the responsibility of the manufacturer, and appears on the market as his/her product. Personalized products are definitively characterized by the production of a one-of-a-kind item. Therefore, AM is preferred and is in many cases the only way of production.

Passive personalization is closely linked to medical devices and products from the field of human-machine interaction (not exclusively in the medical field). Medical products are, for example, implants, epitheses, orthoses, hearing aids, and all affiliated medical devices. All applications have in common that the required 3D data sets have to be obtained from the patient by medical imaging technologies, such as CT (computer tomography) or US (ultrasonography). Based on these technologies and supported by special software, the data set of the personalized part is generated. Then the part is manufactured by AM.

A technically and economically outstanding example is the design and production of hearing aid shells. The process starts off with manually taken an imprint of the auditory canal, which is scanned afterwards to get a digital representation of its 3D structure (Figure 5.21, left). Increasingly, the geometry of the auditory canal is digitally scanned directly and transmitted into the software.

Hearing aids need interior channels, which are important for the necessary ventilation and the tuning of the resonance chamber. The traditional manufacturing of these channels was restricted to straight, drilled channels. By means of additive manufacturing individually adapted, arbitrarily shaped channels are not an issue anymore. Special software optimizes the shell geometry and its interior structure, based on audiometer measurements of the customer. As examples, the interior 3D resonance channel and the ventilation channel are shown in Figure 5.21 (center and right).

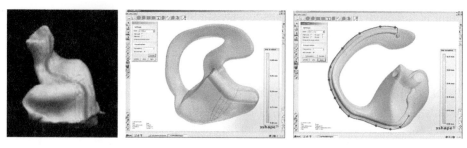

Figure 5.21 Customized production; personalization: design of a personalized hearing aid shell; manually taken imprint (during subsequent scanning) (left); design of the ventilation channel (center); design of the resonance channel (right) (Source: M. Klare [5])

The production is only possible by means of additive manufacturing. Figure 5.22 shows the build platform with some hearing aid shells after the build process and during their removal from the machine.

Figure 5.22 Customized production; personalization: AM production of hearing aid shells (left); hearing aid shells (right) (Source: 3D Systems)

The preferred AM process is the polymerization process. Usually several pairs of hearing aid shells (in this example up to 50) can be manufactured on one build platform. Of course, all shells are different from each other (Figure 5.22, right).

Further applications of personalized products arise in the dental field, such as the bridge in Figure 1.12 (left). Generally, dental elements made by means of additive manufacturing are gaining more and more commercial importance. With the introduction of the completely digitalized process chain, the dental technician becomes – in cooperation with the dentist – an active part of the design and the production process of dentures. The dentist records the teeth geometry by means of a (intraoral) scanner. Supported by special software, the dental technician carries out the dental restoration directly on the computer. The generated dental element is directly manufactured by AM. Compared to this direct design and manufacturing chain, the manual manufacture in low-wage countries will no longer have a competitive advantage.

Figure 5.23 shows the digital design of another kind of dental prosthesis, a removable partial denture (left) that is "called casted model". It concerns a fixation of artificial teeth (not displayed). It has to be detailed with a solid fixation by the remaining teeth and has to resist high loads during chewing as well as during fit and removal.

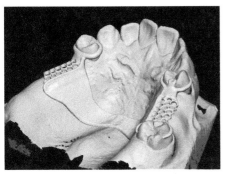 

Figure 5.23 Casted model of a denture: digital design (left); finished part, made by selective laser melting (right)
(Sources: SensAble Dental Lab System (left)/GoetheLab, University of Applied Sciences, Aachen)

The still-applied classical method today is casting. This, however, causes many problems, mainly due to pores and distortion. The direct manufacturing by additive manufacturing processes like selective laser melting (SLM) promises to be a good alternative manufacturing method. A directly printed partial denture made from a CrCo alloy by SLM is shown in Figure 5.23 (right).

Active personalization integrates the customer and his/her creative potential. As most of the customers are not familiar with 3D CAD, the easiest solution is the download of data sets from the internet (e.g., from Shapeways). Of course, these 3D part libraries only provide limited freedom of design. A further step to individualized products leads to the use of websites that support the individual modification of prepared data sets. Moreover, these websites offer the exchange of data sets and some of them also offer tutorials to operate simple 3D programs like Google SketchUp. With these aids users can transmit their ideas into applicable data sets that can be processed by any AM machine.

A similar, but even for non-skilled people applicable, process is "meta-design". Meta-design provides a completely elaborated product, yet allows the customer to create a unique, individual product by varying some key parameters.

As examples, two screenshots of an online personalization of jewelry (ring) are shown in Figure 5.24. By means of a web interface and controllable parameters (scroll bar) the shape of the ring can be adjusted individually. Furthermore, a non-skilled person immediately recognizes the result of her intervention into the design process. After the design process is finalized, the jewelry is produced as a one-of-a-kind part by means of a suitable AM process. The function of the design program usually lies within the responsibility of the manufacturer, while the user or customer has the responsibility for the individual layout.

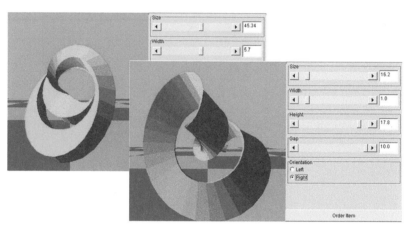

Figure 5.24 Customized production, personalization: meta-designs, software-based personal-ized jewelry design
(Source: ETH Zürich)

5.2.2 Personal Production

If customization, i.e. design and manufacturing, is carried out under the exclusive responsibility of the user or customer, it is called *self-customization*. Usually the design is carried out by the customer, applying either – as mentioned in Section 3.2.1 "Personal Printers" – internet-based CAD tools or, depending on the custo-mer's design capacity, 3D CAD software. Self-customization includes manufacture by means of additive manufacturing, which was up to around 2005 economically not feasible.

Starting at that time, inexpensive, easy-to-operate machines entered the market (see Section 1.3 "Classification of Machines for Additive Manufacturing"). They are called "fabbers" or "desktop printers", but today they are named almost exclusively as "personal printers".

Along with the fast-growing number of personal 3D printers, an enormous growth of internet blogs has occurred. "Fab-at-Home" or "RepRap.org (RepRap-Wiki)", to name just two, initiate and support a fabber movement that can only be compared with the introduction of personal computers 40 years ago, which were based on simple platforms, like ATARI.

An example is the fully functional "Quadrocopter" (Figure 5.25) that was designed and manufactured by its users. Finally, 95% of the required elements were printed on a personal printer from ABS.

Figure 5.25 Self-customization: Quadrocopter "Firefly", consisting of 95% of 3D-printed elements
(Source: *www.Firefly1504.com*)

Figure 5.26 shows a 20 cm (8 inch) high "Darth Vader Buddha" sculpture, printed from PLA. The STL data set for this sculpture is available under a Creative Commons license on the content platform Thingiverse for free download and can be printed at home by a personal printer.

Figure 5.26 Self-customization "Darth Vader Buddha", printed on a personal printer
(Source: data set: Creative Commons license, Thingiverse; picture and printing: GoetheLab, University of Applied Sciences, Aachen)

Figure 5.27 shows an orthosis, a wrist-band, which can be adapted online by a "customize-app", saved as an STL data set, to be printed directly at home and finally be shaped using a hot air gun.

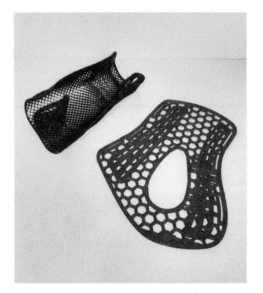 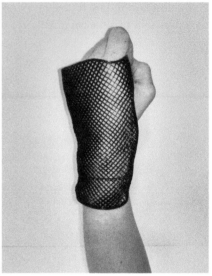

Figure 5.27 Self-customization: orthosis "wrist-band" printed on a personal printer; fore-ground (left). After printing: shaped by hot air gun (leftmost and right) (Source: data set: Creative Commons license, Thingiverse; picture and printing: GoetheLab, University of Applied Sciences, Aachen)

5.2.3 Distributed Individualized Production

Today, an enormous number (more than 300 different models) of personal printers is available. Each fabber belongs to a user, who is familiar with this particular machine, related computers, and the World Wide Web (www). All fabbers and the computers controlling these fabbers can be operated via the internet easily. Thus, a worldwide production net is formed, basically open for each internet user, where he or she can act as internet or cyber producer. Following the idea of the co-working movement (mutual working in variable and mainly temporary spaces) this scenario is called *coproducting*. Building upon a cloud-based organization it is called *cloud production*.

Additive manufacturing is the technology that enables the realization of this kind of worldwide distributed manufacturing net. The different AM machines may be operated independently from each other at any time, if required complementing each other with regards to technical capacity and local availability. The most important issue is that all AM processes are operated by STL (or AMF) files, while all other (traditionally) digitally controlled machines (e.g., milling machines or laser cutting machines) require individually adapted pre-processors. As the large variety of AM machines allows processing of nearly every material and a wide range of dimensions, a coproduction network can manufacture nearly everything, everywhere and at any time.

Since the majority of the owners of personal 3D printers also to a certain extent have knowledge in the field of CAD design or handling of such files, a worldwide design and production net is already developing. It is important to state that such a network is a self-organizing movement, irrespective of whether established structures would like it or not.

■ 5.3 Conclusions

Additive manufacturing allows the fabrication of products with new properties. Additive manufacturing processes not only offer the option to generate parts with very complex geometries and integrated functions that cannot be manufactured without using AM processes (i.e., only by applying traditional manufacturing processes), but AM also allows the manufacture of these products from different materials, and even the variation of the material and its properties within the product.

Moreover, additive manufacturing totally abrogates many conventional rules of manufacturing technology. Additive manufacturing processes allow production of any number of parts of any geometry and from all conceivable materials (at least in the future) at any place in the world. No product specific tools have to be used. Therefore, AM is the basis for individualized and personalized mass production. Additive manufacturing processes not only open up new perspectives for producers, they also allow customers to mesh into the product development process, thus changing the classical structures.

AM marks the change from today's mass production of identical parts to future mass production of different parts in arbitrary numbers of pieces or batches. This is the revolutionary approach with regards to manufacturing processes.

Additive manufacturing supports the idea of producing jointly in an international network, where the local active production community will become part of a cloud-production movement. This is the revolutionary approach with regards to production technology.

■ 5.4 Questions

1. **Why can parts with almost unlimited complex geometry be manufactured by means of AM?**

   As each conceivable part (at least virtually) can be cut, each part can be digitally sliced in layers, these layers can be manufactured, and the part be composed layer by layer.

2. **Which AM processes are able to manufacture parts from materials with varying properties within one part?**

   All AM processes that generate parts voxel by voxel can in principle use materials with different properties for each voxel. Today, these are preferably printing processes with corresponding controlled print heads.

3. **Name at least three examples of material properties that can be varied within one part.**

   Color (3D printing, drop on powder), elasticity (polymer jetting), the type of material (dual extrusion).

4. **Why can AM processes deliver any number of individual items?**

   AM builds the parts layer by layer, regardless of whether the parts are identical or not.

5. **What limits traditional processes with respect to manufacturing of individual parts?**

   Processes not working according to AM need tools. The use of tools demands for a comparably large number of identical items to recover the high costs for the tools.

6. **Why does individualization not depend on the selected AM process?**

   Individualization is a strategy to meet the customer's requirements. It is basically a design method that first of all defines the product. The AM process turns it into reality.

7. **Why is the realization of an AM-based worldwide networked, but locally established, production more probable than of a technology that is based on other digitally controlled manufacturing processes (such as mechanical processing)?**

   Because all AM machines worldwide can be operated with the same type of STL (or AMF) data sets, while most CNC programs require machine-dependent pre-processing.

8. **What are the characteristics of self-customization?**

Self-customization means design and production of a personalized object by means of customer-operated AM devices, predominantly by personal 3D printers.

9. **Which criteria distinguish individualized from personalized products?**

Individualized products aim to target groups of customers. They are manufactured in small series, but still in series production. Depending on the economical break-even point, either traditional (not AM-based) or AM-based manufacturing processes can be used. Personalized products are designed to fulfill the requirements of a single customer. This definitely presupposes an exclusive single item, manufactured by means of AM.

10. **How can movable joints be manufactured in one piece by AM?**

Joints, e.g. hinges, need a defined gap between axis and bush to achieve the desired movability. Gaps are carried out as an uncured layer of resin or as a thin layer of powder or support material. After the build process the layer of powder will be blown off, or the layer of support material will be removed using solvents, thus providing the necessary gap for the joints.

References

[1] Gebhardt, A.; Hötter, J.-St.: *Additive Manufacturing: 3D Printing for Prototyping and Manufacturing*, Carl Hanser Verlag, Munich, 2016

[2] Brücker, Ch., Schröder, W.: Flow visualization in a model of the bronchial tree in the human lung airways via 3-D PIV. Pacific Symposium Flow Visualization and Image Processing, June 3–5, 2003; Chamonix, France

[3] Lenz, J., Shellabear, M.: e-Manufacturing mit Laser-Sintern bis zur Serienfertigung und darüber hinaus. VDI Wissensforum: Rapid Manufacturing. March 1-2, 2005, Aachen, Germany; English Equivalent: Shellabear, M., Lenz, J., Junior, V.: e-Manufacturing with Laser-Sintering – to Series Production and Beyond, LANE, Erlangen, Germany, September 2004

[4] Guidot, R.: *Industrial Design Techniques and Materials*, Flammarion, Paris, 2006

[5] Klare, M., Altmann, R.: Rapid Manufacturing in der Hörgeräteindustrie. In: RTejournal – Forum für Rapid Technologie, Issue 2, (2005), May 2005, ISSN 1614-0923, URN urn:nbn:de:0009-2-1049, URL: *https://www.rtejournal.de/ausgabe2/104*

6 Materials and Design

In this chapter available materials for additive manufacturing as well as design rules and corresponding parameters for the production of quality parts are discussed. Firstly, the different types of materials that can be used for different additive manufacturing processes are discussed. Based on this, the influence of various build processes is presented with regard to the build parameters. Typical applications are illustrated.

As for manufacturing processes in general, it is also important for additive manufacturing processes that the user applies certain design rules. The most important ones, especially valid for AM, are presented here. As this field of work is relatively young, it is still under development. Nevertheless, some basis design rules already have been defined, with whose help the manufacturing and application of additively manufactured parts can proceed in an optimal way. Finally, the selection of a suitable AM process is addressed and advice on quality assurance is given.

This chapter helps to identify and discusses the main influences and peculiarities resulting in parts that significantly differ from traditionally machined ones, mostly with regard to design, manufacturing, and materials.

The explanations provide for the reader and user an understanding of possible issues that may occur in relation to the production of high-quality parts. The discussion of processes and machines was already held in Chapter 2 "Additive Manufacturing Processes/3D Printing", but this chapter supplements that with information from a different point of view. The explanations are exemplary and therefore they do not claim to be complete.

■ 6.1 Materials

In this section selected material-related issues are addressed that have gained more importance by the increasing application of additive manufacturing processes. In contrast to most conventional processes, the material properties of an additively manufactured part are partly determined by the properties of the raw material, but also by the build parameters. This is the reason why the final part properties are the result of the process rather than solely of the composition of the raw material. Influencing factors are not considered separately but analyzed and discussed in the overall context. Table 6.1 shows a selection of material suppliers, structured by different kinds of prefabrication such as filament, powder, or resin/wax.

Table 6.1 Material Supplier for Additive Manufacturing

Supplier	Address (Headquarters)	URL
Supplier: Filament		
3D Systems	Rock Hill, SC, USA	www.3dsystems.com
Advanc3D	Hamburg, Germany	www.advanc3dmaterials.com
Afinia	Chanhassen, MS, USA	www.afinia.com
Avante Technology	Cheyenne, WY, USA	www.avante-technology.com
Avistron	Bergheim, Germany	www.avistron.de
ColorFabb	Venlo, Netherlands	www.colorfabb.com
Das Filament	Braunschweig, Germany	www.dasfilament.de
Doodle3D	Utrecht, Netherlands	www.doodle3d.com
extrudr	Fußach, Austria	www.extrudr.com
fillamentum	Hulin, Czech Republic	www.fillamentum.com
Formfutura	Nijmegen, Netherlands	www.formfutura.com
German RepRap	Feldkirchen, Germany	www.germanreprap.com
igus	Cologne, Germany	www.igus.de
Innofil3D	Emmen, Netherlands	www.innofiel3d.com
MadeSolid	Emeryville, CA, USA	www.madesolid.com
Makerbot	Brooklyn, NY, USA	www.makerbot.com
Material4Print	Paderborn, Germany	www.material4print.de
NinjaFlex	Manheim, PA, USA	www.ninjaflex3d.com
Orbi-Tech	Leichlingen, Germany	www.orbi-tech.de
Stick Filament	Ancona, Italy	www.stickandfilament.de
Stratasys	Eden Prairie, MN, USA	www.stratasys.com
Taulman3D	Saint Peters, MO, USA	www.taulman3d.com
Ultimaker	Geldermalsen, Netherlands	www.ultimaker.com
Verbatim	Egham, Surrey, UK	www.verbatim.de
Volaprint	Gaggenau, Germany	www.volaprint.de
Zortrax	Olsztyn, Poland	www.zortrax.com

Supplier	Address (Headquarters)	URL
Supplier: Powder (Metal and Plastics)		
3D Systems	Rock Hill, SC, USA	*www.3dsystems.com*
Advanc3D	Hamburg, Germany	*www.advanc3dmaterials.com*
Concept Laser	Lichtenfels, Germany	*www.concept-laser.de*
CRP Technology	Modena, Italy	*www.crptechnology.com*
EOS	Munich, Germany	*www.eos.info*
ExcelTec	Lyon, France	*www.exceltec.eud*
Heraeus	Hanau, Germany	*www.heraeus.com*
Praxair	Danbury, CT, USA	*www.praxair.com*
Taulman3D	Saint Peters, MO, USA	*www.taulman3d.com*
Voxeljet	Friedberg, Bavaria, Germany	*www.voxeljet.com*
Supplier: Resin and Wax		
3D Systems	Rock Hill, SC, USA	*www.3dsystems.com*
ARCAM AB (publ.)	Mölndal, Sweden	*www.arcam.com*
DETAX	Ettlingen, Germany	*www.detax.de*
Formlabs	Somerville, MA, USA	*www.formlabs.com*
Lithoz	Vienna, Austria	*www.lithoz.com*
MadeSolid	Emeryville, CA, USA	*www.madesolid.com*
Prodways	Les Mureaux, France	*www.prodways.com*
Stratasys	Eden Prairie, MN, USA	*www.stratasys.com*

6.1.1 Anisotropic Properties

In general, discussions regarding material properties always assume that the parts show isotropic behavior. Isotropic means that the material shows constant characteristic properties in each direction and identical characteristic data at each point of a volume unit. The traditional production, based on the application of tools, presupposes isotropic material behavior, which forms the calculation basis for the engineering.

If a part is manufactured layer by layer, it is not surprising that this part shows discernable differences over its cross section. Therefore, in such a case the material behavior is called anisotropic. This means that the mechanical-technological properties differ within the cross section and perpendicular to it. The layer building method of additive manufacturing processes therefore produces anisotropic parts. The degree of anisotropy may vary from hardly discernible to an extent that has significant influence with regard to the part's stability. Because the degree of anisotropy mainly depends of the additive manufacturing process, which means with regard to the parameters used in the AM process, the orientation of the part in the build space and the principle of its construction play a major role.

The effect of anisotropy can be compensated by changing the orientation in the build space. As a construction rule, the area of highest stress should be arranged in parallel to the build platform. In practice, changing the part's orientation in one area also effects a change of orientation in all other areas. Therefore, a change of the part's orientation should be considered carefully.

Moreover, the effect of anisotropy is closely linked to the method, how adjacent layers are connected. The worst imaginable case is delamination of layers. Obviously, this can appear with FLM procedures, but delamination can also appear with other AM processes. Figure 6.1 demonstrates delamination in a laser-sintered part that was deliberately manufactured with inadequate build parameters to show this effect. As can be seen, the effect varies within the part from local delamination in the compact part to clearly visible single layers.

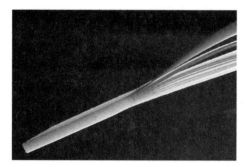 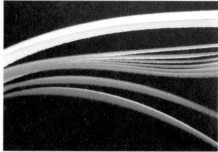

Figure 6.1 Anisotropy: Delamination of layers due to deliberately inaccurately set build parameters; laser sintering, polyamide
(Source: CP-GmbH)

In similar way, but more intensively, anisotropic effects occur within the additive process of powder-binder printing. Often, the porosity of parts has a larger influence on the part's properties than the anisotropy. The negative effect of porosity can be reduced by infiltration, but not eliminated.

Compared with stereolithography, extrusion processes (fused depositing modeling) tend to anisotropic behavior to a higher degree. The process requires a pasty material during extrusion, but does not allow a total melting process, as otherwise the geometrical stability of the growing part would be endangered. Between the single extruded cords, a kind of connecting seam develops that substantially causes the anisotropy. Anisotropy can be reduced by improved extrusion technology; in the first line by suitable heat treatment inside of the machine and the application of thin layers. Professional machines, as well as production machines, achieve a lower-anisotropy of the parts in comparison to personal 3D printers and fabbers, although both types work according to the same principle. With regard to these processes the maximum allowable load in the build plane (horizontal) should be limited.

Regarding process technology, layer-laminate processes show the most distinct anisotropic behavior, since the prefabricated and thus isotropic layers are bonded by binders or glues with completely different properties.

Unfortunately, these rules are not valid for all variations of additive processes. While the rules typically are correct, the processing of plastics and laser sintering of metals (SLM) show a different material behavior. The metal powder is completely molten. This is underlined by the term *selective laser melting* (SLM) for the metal variation of the sintering process. This results in totally dense (> 99%) parts, showing merely low anisotropic effects. In this case rather the kind of material structure is important, depending substantially on the process management. The layers remain visible under microscopic analysis of a polished section.

As already stated, the layer-laminate process fundamentally shows the most distinct anisotropic behavior. The effect depends to a high degree on the process and the material. If paper is used, bonded by gluing, the part is distinctly anisotropic. The anisotropy is reduced if films from plastics are used and bonded by solvents. In contrast, the layers are totally invisible, even in microscopic investigations of microsections, if metal foils are used for the production of solid metal parts by means of diffusion welding or ultrasonic welding. Ultimately, the user should utilize the above information for orientation, but should check the application with the machine's manufacturer or its service department before the intended process application is finally determined.

With regard to direct manufacturing processes, anisotropic behavior of the parts has to be compensated for in the design phase and while fixing the build parameters for the additive manufacturing process.

This requires knowledge about build-direction-dependent deviations of the material properties. In many cases this information is not easy to obtain. Therefore, the suitable design of the parts is subject to experience or requires pre-tests.

A systematic approach, called *part property management* (PPM) – see Section 6.3 "Selection Criteria and Process Organization" – was developed by EOS [1]. The system integrates special material data sets, the so-called *part property profiles* (PPPs). If required, material properties for the build area (x-y) and in the vertical direction (z) are provided.

6.1.2 Isotropic Basic Materials

Today, AM enables the processing of all kinds of material, including plastics, metals, and ceramics. This is valid for all machines of the five AM families (see Chapter 2 "Additive Manufacturing Processes/3D Printing"), although the intensity of usage significantly varies in practice. Sintering of plastics and metals as well as filament-based extrusion can be considered as widespread standard processes, while processes using ceramic or metallic nanoparticles as fillers – mainly in the field of extrusion and polymerization – to a large extent are still under development.

In comparison to materials applied in non-additive processes, the variety of different materials for additive manufacturing and within each material class is still very limited, although it has increased significantly the during recent years, and due to international research is growing continuously. The reason for the limitations is that in most cases a simultaneous development regarding materials as well as the process chains had to be carried out. Materials for laser sintering of plastics, for example, not only have to be able to be melted locally and be solidified in a defined way, but also to be compatible with post-processing mechanisms such as, for example, coating or grinding that might lead to rounded edges.

Chemical additives and process-related details such as the kind of shielding gas and the gas distribution, preheating, suppression of local evaporation, oxidation, and other process-internal effects, as well as emissions, have to be considered. In this list of requirements lie also the reasons why from a chemical point of view PA materials for injection molding or fluidized bed coating are highly similar, but these cannot be used for additive manufacturing without modifications.

Therefore, materials for additive manufacturing usually are developed by machine manufacturers or under their responsibility. These manufacturers declare the material as a proprietary development and sell it to their machine customers exclusively.

The steady increase in total material consumption favors the activities of "third-party suppliers", who have already entered the AM business. Mainly in the field of plastic materials for extrusion, laser sintering, and stereolithography, independent markets are expanding.

Metal powders are very similar to standard powders for laser coating and laser weld overlay and therefore well established in the market for many years. The user can choose from a wide range but has to qualify the process in most of cases, i.e. to provide a material data sheet is the user's own responsibility. Alternatively, materials can be used that have been released by the machine manufacturers. As a consequence, the low number of available materials and their price level have to be accepted.

Some AM-related issues are up to now hardly investigated or are not yet in the focus of users. The largest group of issues is related to long-term behavior, which with regard to the application of additively manufactured parts in the field of *rapid prototyping* is not of great significance but becomes very important regarding the application of *rapid manufacturing.*

The most important issues include aging, UV stability of plastic parts, corrosion, decomposition, sedimentation, and oxidation for metal powders, as well as pores and inclusions for all additive processes. In the scope of this book not all issues can be discussed in detail.

6.1.2.1 Plastics

Plastics were the first group of materials to be processed by means of additive manufacturing processes and still represent the largest portion of materials used for additive manufacturing. Materials for laser stereolithography are acrylates or epoxy resins, which have to be suitable for polymerization. Instead of the sticky, unpleasant-smelling, and brittle materials of the early 1990s, today materials are available that behave like materials for plastics injection molding. This was achieved by chemical modifications and by addition of nanoparticles, which significantly improved the thermal and the mechanical behavior and the technological parameters that can be obtained. Moreover, the range of materials has increased, so that transparent and translucent, elastic, rigid, and other types of materials are now available.

The manufacturer Formlabs, for example, has introduced into the market a new bio-compatible resin for dental applications (see Figure 6.2). This certified resin is the first bio-compatible SLA material for a desktop 3D printer; see Section 3.2.1 "Personal Printers".

Figure 6.2 Application at the textures of a 3D printed object with the "Dental SG-Filament" of Formlabs
(Source: Formlabs)

For laser sintering of plastics, polyamide (PA) is the preferred material. Polyamides belong to one of the most popular thermoplastic material families for injection molding. Nevertheless, or probably for that reason, problems appear, because polyamides for additive manufacturing differ significantly from those used for plastics injection molding, even if they are chemically identical. This on the one hand is caused by the thermal process but on the other hand by the large number of variants. A material that is completely molten in the process and injected under high pressure in a tool and then cooled down shows different properties to the same material that is locally molten under atmospheric pressure, deposited layer by layer, and partially solidified by heat conduction.

"Polyamides" is also the overall term for a very big family of plastics with various special properties. Therefore, the term "polyamide" is not sufficient to clearly determine the material. Industrial products are typically made from polyamide 6 or 6.6, while for laser sintering preferably polyamide 11 or 12 is used. Besides its low tendency for water absorption, polyamide 12 especially is used for reproducible production conditions due to its large process window. The powder particles are of a grain size of mainly between 20 and 50 µm.

Warpage and distortions of single layers or of entire parts was a big issue at the beginning of laser sintering. Today, these defects are reduced to a minimum by preheating and due to improved scanning strategies.

For laser sintering, a relatively broad and continuously increasing range of materials is available. This includes flame retardant, elastic, or aluminum (nano powder) filled grades as well as those for medical applications that can be sterilized and used as a permanent implant. Improved mechanical properties are achieved with glass-filled powder. This term is misleading, because instead of the expected glass fibers (as are used for injection molding), spherical and rice-grain-shaped particles are used. Powders of this kind provide in comparison to unfilled powders higher stability, but do not achieve the properties expected from materials that are filled with glass fibers and processed by injection molding. Parallel to the increasing number of worldwide installed AM systems, an independent third-party market for powder materials is developing, which influences the economic situation as well as the qualification of new application-driven grades.

Regarding industrial application, widespread powder types like polyamide 6.6 are available, but not often used for AM yet.

Although the development and manufacture of new products strongly depends on the availability of high-performance plastics, at present only a few are available. A PPSU (polyphenylsulfone) material is available for Stratasys's extrusion processes, and since the end of 2010, the high-temperature material PEEK (polyetheretherketone) has been established by EOS as PEEK HP3. PEEK has excellent heat- and corrosion-resistant properties. The material is flame and heat resistant, chemically resistant, has a high mechanical stability, a low weight, is bio-compatible, and can be sterilized. It has a melting point of 334 °C and requires a processing temperature window between 350 and 380 °C. This is far above the temperature ranges of actual laser sintering machines processing plastics and set off the development of a completely new high-temperature machine, the EOS P800 (Table 3.7).

The company Indmatec GmbH was the first to successfully make PEEK printable with the *fused filament fabrication* (FFF) technology. The filament "PEEK MedTec" can be used for various medical, dental, and surgical applications, due to successful biological qualification, according to EN ISO 10993-5/-4/-18. Moreover, the filament can be used for prototyping of medical products, like prostheses.

Another family of materials for laser sintering comprises polyamide-coated particles from nearly any materials. The most important application is the use of coated foundry sands for sintering of cores and models for sand-casting. The particle size of the uncoated material is about 50 μm.

Similar materials are available as metal sheets or foils. In this case the coating acts as a binder, making AM to a two-stage process. This kind of process, however, is less widespread, as for most of the applications a one-stage process is available. Nevertheless, it opens up new options to qualify materials by means of the binder, and moreover to use filled particles, acting like granules.

Amorphous polystyrene is also used as material for laser sintering, predominantly for manufacturing of lost cores and cavities for the *lost foam process.*

For extrusion processes, especially for FDM applications, but also for PolyJet processes, proprietary materials are available, which are commercialized by the manufacturers (see Chapter 2 "Additive Manufacturing Processes/3D Printing").

With regard to polymerization it should be noted that lamp-supported processes, i.e. including polymer printing, work with acrylates, while for laser stereolithography epoxy resin is used. Epoxy resins achieve better part properties, but need more energy for the polymerization and therefore require the use of lasers.

The basic material for FDM is an ABS plastic of higher quality. Since ABS is often used for plastic injection molding in series production, it frequently is considered as series material. Here the above-made remarks with regard to the influence of nominally identical materials on the process are valid. The user should be aware that ABS is a standard polymer, which is much more resistant against lower temperatures than is polyamide.

In all AM processes that take the build material out of a separate storage bin (PolyJet, FDM), instead of storing it in the build chamber (laser sintering, laser stereolithography), the material and consequently the part can be colored. Parts manufactured by laser sintering or stereolithography can also be colored; however, in that case the complete material, stored in the build chamber, has to be dyed. Therefore, only parts in this chosen color can be manufactured.

To summarize, it has to be stated that an increasing variety of plastics can be processed by means of additive manufacturing. This is valid for all five AM families, described in Chapter 2 "Additive Manufacturing Processes/3D Printing". In Figure 6.3 the plastics processes are shown in the traditional "plastics triangle". It differentiates the various plastics with regard to their basic structure (amorphous, crystalline) and their temperature resistance. Figure 6.3 should only be used for orientation purposes. In particular temperatures should not be taken from the diagram but be confirmed by the supplier.

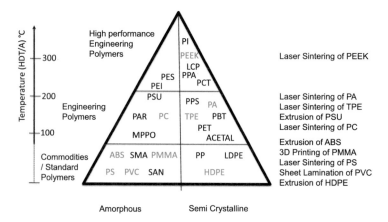

Figure 6.3 "Plastics triangle" with plotted AM processes and related materials (based on [2])

Today on each application level an AM material and a corresponding AM process are available. Only the polyimides are an exception. However, polyimides comprise a highly interesting polymer material family, showing high stability, heat resistance, and chemical resistance. Therefore, they are highly desirable printing materials.

6.1.2.2 Metals

The most frequently used additive process to produce metal parts is sintering in the variation "selective laser melting (SLM)/Powder Bed Fusion (PBF)" (Chapter 2 "Additive Manufacturing Processes/3D Printing"). The material is supplied as powder with an initial grain size in the range of 20 to 40 µm. Since the diameter of the laser beam, the layer thickness, and width of the track are in the same range, the scanning structure is clearly visible on the surface (Figure 6.4).

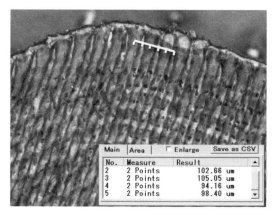

Figure 6.4 Upper layer of a metal part, top view; manufactured by laser sintering (SLM)/powder bed fusion (PBF).
(Source: GoetheLab, University of Applied Sciences, Aachen)

The materials highly resemble those that are used for laser coating or laser welding with filler material. Therefore, a wide range of grades from different suppliers is available, for which expert knowledge on a high level has already been collected. Usually, powders supplied by the manufacturer of the AM machine are delivered with material data sheets that are based on reliable parameters, also including optimized parameter sets.

Nevertheless, the necessity to qualify or at least to evaluate material in house should be considered.

For AM metal sintering processes (SLM), stainless steel, tool steel, CoCr alloys, titanium, magnesium, aluminum, and precious metals like gold and silver are available. The first objects manufactured from copper were presented already in 2011 [3], but the process was not yet developed to a marketable level. Proprietary materials were developed particularly for dental applications and are often commercialized as a package with specialized software and modified machines. Figure 6.5 provides an overview of the different metals for AM that are processable by means of laser sintering (laser melting).

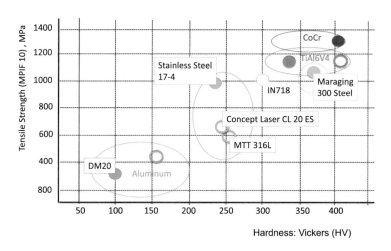

Figure 6.5 Selection of materials for AM metal processes, based on EOS, 3TRPD, Concept Laser, and ref. [4]

The data were issued by the service bureau 3 T RPD Ltd. and are based on EOS information (filled circles). In addition, information of other suppliers – mainly from ref. [4] – has been added (void filled circles).

It is important to mention that other suppliers offer similar materials in a comparable range.

The layer laminate process (LLM) works predominantly with sheets or films from plastics or paper. Metal parts can be manufactured by means of the *ultrasonic consolidation process* (Fabrisonic). It works with aluminum strips that are joined to the surface of the partly finished object by ultrasonic welding. Additionally, the machine is equipped with a 3-axis milling device that generates the contour of the layer in the same clamping position. The process provides completely dense aluminum parts. As the part remains cold (room temperature) while produced, even electronic elements can be inserted into integrated pockets, and be covered and sealed by the subsequent layer. The parts are predominantly used as integrated sensor housings in aeronautical and deep-sea applications.

In summary, it can be pointed out that the range of metal materials for additive manufacturing is larger, in comparison to plastic materials. Moreover, the metal part's properties resemble those of traditional manufacturing processes more than for the application of plastics.

A compilation of applications of AM for the manufacture of micro-parts and dental objects, as well as the processing of various metals by means of the M3D process (Optomec), was published under [5].

6.1.2.3 Ceramic Materials

Additive manufacturing processes processing ceramic materials still form a special niche regarding layer manufacturing processes. Although for each of the five AM families at least one process is available, ceramic 3D printing applications are still rare.

Often, the powder-binder process (also known as 3D printing) is used, providing the possibility to work with different powder-binder combinations, including proprietary grades. For these applications a two-stage process has to be accepted and qualification of the material has to be carried out on one's own authority.

Also, *selective laser melting* (SLM)/powder bed fusion (PBF) is used, to process ceramics and different types of glass [6].

A laser-stereolithography process that uses a resin with ceramic filler and delivers very detailed ceramic parts was developed and commercialized by the company Lithoz in cooperation with the university of Vienna (Professor Stampfl).

Layer laminate manufacturing (LLM) basically is suitable to process ceramic foils. The so-called CAMLEM process resembles the traditional "tape casting" and is a two-stage process with additional sintering.

Today (2018) the materials cover the whole spectrum of ceramics: aluminum oxide or alumina, Al_2O_3; silicon dioxide or silica, SiO_2; zirconium oxide or zirconia ZrO_2; silicon carbide, SiC; and silicon nitride, Si_3N_4.

The products are monolithic ceramics, predominantly including channels for the manufacture of high-temperature elements, like heat exchangers. Defined macro-pores, supporting the ingrowth of implants, are particularly important for the application of resorbable bio-ceramics. Micro-pores allow the manufacture of micro-reactors. For a detailed overview see ref. [7].

6.1.2.4 Composite Materials

In contrast to layered carbon fiber production, regarding AM it is common to name all materials "composite materials" that are composed from more than one material grade. The composite materials used for AM are either prefabricated filaments or composed within the machine prior to printing. The composition can be achieved from different powders that are processed in an SLM/PBF process or by processing simultaneously with an extruder as applied to print the "Strato" (see Section 2.1.6.3 "Extruding and Milling – Big Area Additive Manufacturing (BAAM)"). The "Strato" material is composed from ABS pellets and short cut or milled carbon fibers. Alternatively, separate filaments can be applied using a dual extruder FDM process with one extrusion path for each material.

In this sense "graded materials" or "tailored materials" (see Section 6.1.3 "Graded Materials and Composite Materials") can be regarded as composite materials as well.

In principle LLM composite elements can be made with integrated fibers or fabrics, as long as the reinforcements are available as prepregs or flat semi-finished foils that can be integrated into the process. A specially adapted process for the manufacture of reinforced curved elements, using ceramic fibers for which cutting is avoided, is described under [8] (see Figure 6.6). This process enables the arrangement of the layers under defined but different angles, to adapt the structure to the expected load.

A general disadvantage of composite materials is the fact that it is mostly very complicated or even impossible to separate the materials after the end of life of the product, in order to achieve a proper recycling.

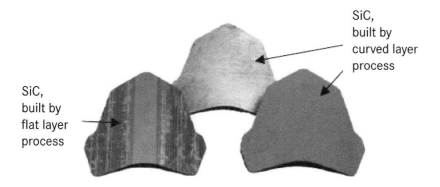

SiC, built by curved layer process

SiC, built by flat layer process

Figure 6.6 Layer laminate manufacturing (LLM): reinforced curved elements with integrated SiC-fibers [8]

6.1.2.5 Further Materials

6.1.2.5.1 Concrete

A new material for additive manufacturing that is currently being researched worldwide is concrete, which opens up a new application for the AM technology with a special focus on house building. Various endeavors are currently underway, mainly in the field of prefabricated concrete construction, to introduce this modern production process into the construction industry. The three central features of this continuously growing technology are speed, accuracy, and profitability of the 3D printing of walls, floors, roofs, plates, and other structural elements. With regard to printed walls, concrete objects with lots of details and precision are possible, as numerous examples now illustrate.

According to a new report of the market research company "MarketsandMarkets" the market for concrete from the 3D printer will grow from US$24.5 million in 2015 to US$56.4 million in 2021.

6.1.2.5.2 Carbon

Carbon is a material that in future will take over many applications of metals. Carbon fiber reinforced parts can bear a load 20 times higher than steel while having only a fraction of the weight of equivalent steel parts. Carbon reinforced parts of the UHM (ultra high modulus) type have a tensile strength up to 4560 MPa (N/mm^2) and a modulus of 395 GPa. 3D printed parts from carbon will find their application in the aerospace, automotive, and armaments industries, and in many other areas. Currently, Volkswagen is investigating an axle made from carbon for the electrically driven Golf 7 (Germany), to compensate for the high weight of the batteries.

The company Markforged has presented the Mark Two, the worldwide first carbon-3D printer. It prints alternately continuous carbon fibers and a plastic carrier-layer. The parts are, above all, very stiff in the x-y-preference direction and therefore highly loadable.

6.1.2.5.3 Food

The further development of processes and processable materials continuously opens up new fields of applications, as in the food industry and the catering trade. Universities, research institutes, and the industry worldwide investigate the diversity of 3D printing technology with regard to foodstuffs, also known as "food printing". The second largest German producer of confectionery, Katjes, for example, developed the "magic candy factory", printing individual *fruit gums*.

Beside fruit gum, foodstuffs like chocolate, marzipan, cakes, pastries, and much more are produced by 3D printing in diverse forms and sizes.

Figure 6.7 "Magic Candy Factory": 3D printer for fruit gum
(Source: Katjes)

Processes applied up to now use the piping bag method. The dough or paste is inserted into the printer in the form of a cartridge and then slowly dosed and layered through a nozzle. Any foodstuff that can be transferred into a paste can be used.

6.1.3 Graded Materials and Composite Materials

Isotropic material behavior is obviously the rarely mentioned basis of all classical design und construction rules. Most current products were developed based on these rules, and not only the engineering design but also the production was optimized accordingly.

However additive manufacturing enables the manufacture of products from materials with locally different properties that can be adapted selectively to the loads occurring during their use. Parts made from such materials cannot be manufactured by traditional manufacturing methods, but are feasible by using AM technology. In printing, the characteristics of such materials are not only defined by the raw material, but also by the type of local material application and solidification, and are thus determined by the process. AM allows one to locally influence and even create the required material for a defined application.

For example, the parameter "color", which also is a material property, can be adjusted during the 3D printing process (powder-binder process) for the production of continuously colored parts. In the future the same process can be applied to adjust the flexibility or other properties of a part. The polymer printing process (Objet) demonstrates the practicability. It allows the processing of different materials in the same build process and even variation of their respective ratio during the process. Parts made from two components, e.g. hard/soft combinations, already today can be made to imitate two-component parts made by injection molding. Multi-material and multi-color processes were presented in Section 5.1.4 "Multi-Material Parts and Graded Materials".

These examples characterize the beginning of the manufacture of anisotropic products with materials that are locally adapted with respect to the loads. The first steps demonstrate the basic principle that will be intensively developed in the future and will allow not only the manufacture of industrial products, but also of food, as well as medicine and artificial organs. Examples are already available but are still in the research and development stage.

All processes that are supplied with materials using small cartridges, containers, or wound-up filaments are able to run in multi-material mode just by omultiplying the material-application units. PolyJet – as well as 3D printing (powder-binder) processes – already commenced to use this technology.

Graded materials and composite materials are not only a challenge for additive manufacturing. To profit from the emerging possibilities, the engineering designer has to be familiar with them. Design rules have to be extended to allow computation of parts from anisotropic materials with suitable material parameters. People are working on essential guidelines mainly to support the advantageous applications of AM, such as freedom of design, accuracy of fit, or tolerance management.

■ 6.2 Construction – Engineering Design

To benefit from the advantages of additive manufacturing, certain design rules have to be observed. These are primarily derived from the practical application of additive manufacturing and are up to now only to a lesser extent the result of methodical investigations. As this field is quite new, completed construction guidelines – as available for casting or milling – are missing. An actual summary can be found in ASTM WK 54856 "New Guide for Principles of Design Rules in Additive Manufacturing".

6.2.1 Tolerances – From the Digital Design to the Part

AM parts have to be manufactured in accordance with the 3D drawing by defining the tool path needed to generate the printed contour. In general, the tool path firstly is defined by the center of the tool, for example the laser beam. To achieve the correct geometry, the tool path has to be corrected, for example by half of the diameter of the laser beam. Thus, the generated contour becomes identical with the designed contour of the part. In laser-based processes this method is called "beam with compensation".

Therefore, the exact contour of the part has to be defined during the design process in the "center of the tolerance" zone of the layers, thus achieving a symmetric tolerance field all over the part. A drill hole with a diameter of 20 mm and a tolerance field of e.g. 0.3 mm has to be designed with a diameter of 20 mm ± 0.15 mm tolerance. If the drill hole would be defined with a diameter of e.g. 20 mm +0.2/−0.1 mm tolerance, which also would result in a tolerance field of 0.3 mm, it would be technically correct, but the outer contour of the element would not be generated exactly. To address this issue, the term "digital to object" was formed.

6.2.2 Design Freedom

The option to generate nearly every conceivable shape or geometry of parts by means of AM is a significant advantage and opens up a variety of design variations. The most important ones are discussed with regard to the design limitations of injection molding and die-casting. Rules for other applications can be easily derived from them.

By using additive manufacturing processes, the part can by manufactured in one piece and without any mold joints. No parting lines have to be defined. Undercuts can be easily realized, without increasing the production costs. Fine details and narrow gaps can be made economically, without electric discharge machining (EDM). No release angles have to be provided and no simulation of the injection molding process has to be done to optimize the manufacturing process and the part's geometry. Also, complicated cooling channels do not cause manufacturing problems.

Although it is stated that larger modifications of wall thicknesses do not cause problems when applying AM, the user should be aware that all additive processes are based on phase changes. Therefore, material agglomeration should be avoided on principle.

6.2.3 Relative Fit

In practice, often exact absolute dimensions of a part consisting of two or more elements, whose relative position to each other, however, is not significant, are required. In such a case an appropriate relative fit is sufficient. To realize this, the elements are positioned in the build chamber face to face and as close to each other as possible, to achieve a gap of 0.1–0.2 mm. This ensures that adjacent elements with exact total length fit to each other, independent of whether a perfect parting line exists or not, and how complex the boundary contours are. Even distortions caused by the process would not affect the accuracy of the overall fit.

This discussion touches the "spare parts" issue. All tool-bound manufacturing methods – i.e. almost all actual manufacturing technologies – are based on the manufacture of interchangeable parts as a precondition for (distributed) mass production. If additive manufacturing processes compete with mass production, also additively manufactured parts can be produced interchangeably with respect to the production of series or spare parts. Basically, however, additively manufactured elements are "one of a kind" and are not designed to compete with those that are conventionally designed. This aspect is discussed further in Chapter 5 "Perspectives and Strategies of Additive Manufacturing".

6.2.4 Flexures, Hinges, and Snap-Fits

The most important design elements made from plastics are snap-fits, living or film hinges, and flexures. They form the basis of articulated elements that are integrated into additively manufactured parts. As a main benefit, products composed of several parts that traditionally require assembly of various elements and a variety of tools for manufacturing, as well as assembly lines and individual alignment and adjustment, can be avoided. Additive manufacturing processes based on plastics allow the manufacture of these elements in one build procedure, simultaneously providing the kinetic function of the parts. This feature is also called a "non-assembly mechanism" [9]. The preferred affiliated processes are polymerization and laser sintering, but also extrusion may be applied. The powder-binder process and the lamination processes are less suitable for the manufacture of this kind of parts.

For snap-fits the same design rules are valid as for parts made by injection molding. As these are made from plastics, they tend toward creep, and should not be loaded while in operation. The wall thicknesses should not be below 0.5 mm and the free moving space should be as large as possible, in order not to overload the parts. Film joints (living hinges) should have a wall thickness of 0.5 mm. Exceptions from this rule should be backed by individual tests.

Regarding laser-based processes, hinges can be created similar to drill holes; however, for bigger diameters between 1 and 10 mm, adapted values have to be applied. Both elements of a hinge can be manufactured by means of AM in an assembled state in one build process, avoiding post-processing, assembly, and adjustment. Also, top layers may be added, to generate a rivet-head type of joint.

Preferred processes are polymerization and laser sintering of plastics. To guarantee the flexibility, sufficient clearance has to be provided between adjacent walls. According to many sources it is sufficient to provide only one or two uncured (or unsintered) layers to achieve a hinge able to work after cleaning. In practice, however, rather twice as many should be applied to avoid malfunctions.

"Flexural hinges", usually called "flexures", are solid-state hinges, used in micromechanics to manufacture ultra-precise positioning systems. Flexures are used to avoid problems with tolerance, clearance, and hysteresis. Using AM, "flexures" are considered as construction elements. With regard to plastics processing, "film hinges" can be considered as "flexures". The integrated high-precision mirror holder and the positioning system mentioned in Chapter 5 "Perspectives and Strategies of Additive Manufacturing" (see Figure 5.5) are good examples for an extended "flexure" design that can only be manufactured by means of AM.

6.2.5 Orientation and Positioning of Parts in the Build Space

Parts manufactured by means of AM do not need clamping. The positioning is either achieved by supports built with the part, or by the material in which the part is embedded during the build. In contrary to most of the traditional manufacturing methods, AM allows the arrangement of the part at any angle on the build platform. This option is used to reduce the stair-stepping effect in important areas of the part and to define preferred loading directions.

As already mentioned, any 3D data set, particularly if using the STL format, can be processed by additive manufacturing. Generally, the parts can be arranged on the build platform in any position. In practice the positioning of the part is determined by the orientation of the layers, which should be positioned in parallel to the build plane. If possible, functional surfaces should be arranged facing up, while surfaces of lower importance and areas connected to supports should be oriented face down. This is also valid for PolyJet processes, because also the up-facing surfaces show better quality in comparison to the down-facing ones.

Below, two examples for the right positioning in the build space are presented, following the Association of German Engineers (VDI) guideline 3405, sheet 3 [10].

With regard to long parts it is of particular importance to orientate the small edges towards the recoating device (Figure 6.8).

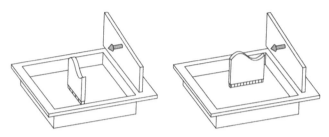

Figure 6.8 Positioning of long parts: left: bad positioning; right: suitable positioning in the
build space
(Source: VDI 3405.3)

To avoid unnecessary or intensive post-processing it is of advantage to orientate
the part shown in Figure 6.11 according to Figure 6.9 (right) in the build space.

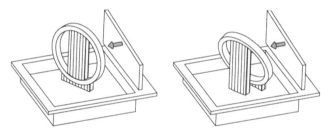

Figure 6.9 Avoiding intensive post-processing: left: bad positioning; right: suitable positioning
in the build space
(Source: VDI 3405.3)

As already mentioned in Chapter 2 "Additive Manufacturing Processes/3D Printing", stair-stepping is a characteristic effect of additive manufacturing processes.
The visibility of steps can be minimized if the angle of inclination between the
z-plane and the y-build plane exceeds 20°. According to this guideline following
ref. [11], flat zones show clearly visible stair-steps, while inclined walls reduce the
effect.

6.2.6 Bores (Holes), Gaps, Pins, and Walls

Holes and other details such as linear and contoured gaps cause some problems.
Due to the "stair-stepping" effect, holes are not perfectly circular, and their shape
depends on their position in the build space. As rule of thumb for laser sintering of
polyamide, a hole diameter should not be below 0.5 mm for it to still be recognizable it as a hole (though not of good quality), provided the wall thickness is 0.3 mm
or less. With a wall thickness of more than 0.6 mm the minimal recognizable hole
diameter increases to 0.7 mm [11]. Test cubes designed by the German computer

magazine "c't" [12], which were manufactured by different service providers, convey an idea about the capacity and limits of established AM processes offered as (internet) services (Figure 6.10).

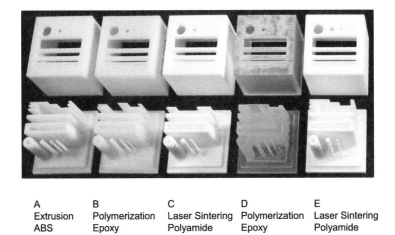

A	B	C	D	E
Extrusion	Polymerization	Laser Sintering	Polymerization	Laser Sintering
ABS	Epoxy	Polyamide	Epoxy	Polyamide

Figure 6.10 Geometrical test cubes for different geometries and AM processes [12]

The cubes have a wall thickness of 2 mm, a base of 3 × 3 cm, and the same height. The base part has free-standing walls and pins of 5, 2, 1, 0.5, and 0.2 mm thickness or diameter. The cover (in the picture at the top) has the fitting slots and holes, when it is put top to bottom on the base part.

A clearance of 0.1 mm allows exact assembly if the parts were made correctly. The parts were manufactured by three different service providers, taken from the internet. The AM processes and machines were not mentioned in detail, as the users only were interested in the results (see Chapter 1 "Basics of 3D Printing Technology"). The orders were placed based on internet information with definitions of accuracy such as "standard" or "fine" as well as verbal material descriptions. In Figure 6.10 five test cubes are shown, which were manufactured from ABS by extrusion (A), epoxy resin by polymerization (B, D), and polyamide by laser sintering (C, E). Figure 6.10 clearly shows that the quality of the parts is not only influenced by the kind of process, but also by the material and the sum of other influences of the process chain.

Holes with a diameter up to 0.2 mm and walls with a thickness below 0.2 mm could not be made at all and for those with a dimension of 0.5 mm, the machine had to be calibrated in advance in order to achieve reliable results. Interestingly, these results are valid for polymerization processes as well as for sintering processes. Even the extrusion of ABS shows similar results with too-large diameters for small pins (Figure 6.10 A).

With all additive manufacturing processes, problems arise regarding the cleaning of very small holes. Sintering tends to leave partly molten particles in the holes or slots (Figure 6.10 C and E), which can be generally be removed by sandblasting; however, sometimes they adhere permanently. Due to processing of liquid material, polymerization processes seem to deliver better quality, although solidified material can block the openings during or after the cleaning procedure (Figure 6.10 B and D).

If holes made by additive manufacturing show "stair-stepping" in parallel to the center line due to the build process, increased wear occurs during the use of the parts, causing the growth of the holes' diameter within a short period. This effect has to be taken into account if precise holes are required. Figure 6.11 shows holes with a diameter of 0.5 mm and integrated internal cooling channels, which can be produced by additive manufacturing.

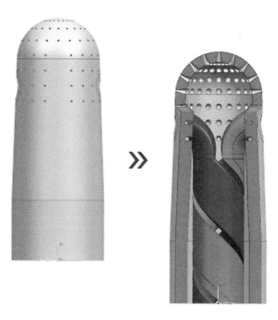

Figure 6.11 Holes with a diameter of 0.5 mm and integrated internal cooling channels (Source: Aachener Zentrum für 3D-Druck (Aachen center for 3D printing)/IwF GmbH)

Sometimes it is of advantage to mark the holes during the build process and to drill the holes mechanically during post-processing, as it is an established procedure within foundries. Also, the manufacture of a drilling template, which is sintered in the same build process, can be a good solution. All remarks regarding holes are basically also valid for gaps.

Free-standing walls are easier to manufacture than pins. It is possible to generate walls with a thickness of 0.5 mm. Sometimes even pins with a diameter of 0.2 mm can be seen, but it is evident that this is touching the limit of resolution. As a prerequisite the process must be calibrated, and all details need to be processed carefully.

All the above-given information presupposes the calibration and professional handling of the production machines. A laser-stereolithography machine that has been calibrated properly can produce parts with channels with a wall distance of 0.05 mm if the parts are exactly adjusted in the center of the build platform, thus minimizing scanner problems.

■ 6.3 Selection Criteria and Process Organization

The selection of a suitable additive manufacturing process depends on the application level of the part (Chapter 1 "Basics of 3D Printing Technology") and differs depending on whether a prototype or a product has to be printed.

If a prototype should be generated, e.g. a solid image or a functional prototype, the selection starts "at the end of the process". First, a material suitable for additive manufacturing has to be chosen that corresponds best with the material properties of the later series part. Second, the machine has to be determined that best fits to the selected material and process. The 3D data set of the part has to be transferred and afterwards the part has to be manufactured. Depending on the chosen material and the selected process a solid image or a functional prototype is generated.

If a product is required, also the material has to be defined in the first instance.

The engineering design of the part has to be based on the properties of the chosen material and the design guidelines for additive manufacturing processes. Other parameters like the orientation of the part in the build space have to be added in the engineering design phase. For the subsequent additive build process, identical parameters to those applied in the engineering design phase have to be employed.

The operator of the machine of the additive manufacturing plant is only responsible for its proper function and professional operation (see Section 3.1 "Data Processing and Process Chains"). To carry out these activities in a reproducible manner, possibly at different machines, the operator needs to have a kind of management system that today is rather based on experience.

With the increasing use of AM,

- the number of build processes,
- the number of repetitions of earlier production runs, which have to achieve consistent quality,
- the number of processable materials, and
- the number of different machines (printers)

also increase.

Under these circumstances the currently still frequently applied "hand to mouth" method regarding management systems can no longer be maintained.

EOS developed and published a management system providing online support for their customers [1, 11]. It is available for sintering of polyamide and in a first approach also for sintering of metals.

An important element is the profile of the part's properties, a database that not only is closely linked with the material properties, but also with the quality management system. The user defines the quality level for a defined material by pre-setting the layer thickness in five steps (60 to 180 microns). Based on this, most subtle details are selected. The profile provides reliable information about dimensioning of laser-sintered parts with regard to mechanical strength, elongation at break, and moduli of elasticity. These properties are separately available for the x-y-plane and the z-direction, thus taking into consideration anisotropic effects.

The system includes design and engineering guidelines as well as parameters according to Section 6.2 "Construction – Engineering Design".

Additionally, comments are given with regard to volume elements, design recommendations for proper cleaning, as well as optimization proposals for economic production, e.g. by means of staggered arrangement or nesting of parts. Different versions are stored to be able to identify data of preceding build processes and to recall them. Special software enables the customers to create their individual data sets, to store them, and to activate them for later applications. As a link to quality management a report is made for each build and stored. Supporting platforms of this kind will become the standard of all suppliers in the foreseeable future.

■ 6.4 Conclusions and Outlook

Additive manufacturing processes are manufacturing technologies, and like other technologies dependent on the careful selection of materials, processes, and the quality of engineering design. Therefore, appropriate engineering guidelines have to be developed and applied. The five additive manufacturing families in Chapter 2

"Additive Manufacturing Processes/3D Printing" show different advantages and disadvantages, which have to be taken into account in the engineering phase. All issues are linked and have to be considered in parallel.

Due to the increasing number of machines and materials, the manufacturing organization cannot be based only on experience any longer. Linked data files and management systems are required, to be able to record additional quality requirements accordingly.

■ 6.5 Questions

1. **Why is anisotropy an important issue regarding additive manufacturing?**

   To manufacture a part layer-wise leads to bonding defects between the layers that depend on the material and the bonding method.

2. **Which additive manufacturing processes for plastics show anisotropic effects and how can the degree of anisotropy be judged relatively?**

   The ranking of manufacturing processes according to part quality ranging from "nearly isotropic" to "anisotropic" is as follows: polymerization, laser sintering, FDM, powder-binder processes (3D printing), and layer laminate manufacturing.

3. **Why does laser melting of metals provide better quality regarding isotropy than laser sintering of plastics?**

   Because by laser melting of metals the material is completely molten.

4. **What are the advantages and disadvantages of qualification of materials on-site?**

   Advantages: well-known material, individual composition and archived build data, low costs.

   Disadvantages: dependence on material supplier with possibly higher costs, missing know-how.

5. **What are typical grain sizes for sintering of polyamide?**

   20 up to 50 microns.

6. **What sorts of plastics can be processed by AM? Plot them in the plastics triangle.**

   See Figure 6.3

7. **How can articulated parts can be constructed in one build? Which geometrical parameters (clearances) have to be taken into account?**

For example, film hinges (living hinges) can be applied. Film hinges do not show clearances. Alternatively, some layers (1 up to 3) of uncured/unsintered material can be applied to provide the clearance needed; magnitude of clearance: 0.3 up to 0.8 mm.

8. **Why there is no clamping issue for the parts produced by additive manufacturing? How are the parts fixed?**

The parts get fixed by supports on the build platform or by the surrounding powder bed. As there is no need to fix the parts, they do not get stressed by fixing forces.

9. **What does "relative fit" mean? Why it is only relevant with regard to additive manufacturing?**

a) It means that the individual dimensions of the elements that form a part are not important, if the absolute dimensions are correct.

b) To achieve this, the parts are manufactured in one build process, positioned opposite to each other with minimal clearance. If the absolute fit is given, the relative fit is not important. Consequently, the parts compensate their interior misalignment if the overall dimensions are correct. This can be done only by additive manufacturing, as it allows addition of material.

References

[1] Mattes, Th., Pfefferkorn, F.: Part Property Management (PPM): Standardization and comparability of building processes and their results. Download from: *http://www.eos.info/en/products/solutions/part-property-management.html*

[2] Kruth, J.-P., Levy, G., Klocke, F., Childs, T. H.C.: Consolidation phenomena in laser and powder-bed based layered manufacturing. *Annals of the CIRP*, Vol. 56, 2/2007

[3] N.N: Lasersintern von Kupferpulver, Photonik 2/2011

[4] Castillo, L.: Study about rapid manufacturing of complex parts of stainless steel and titanium. AIMME and TNO report 2005, Download link: *http://www.rm-platform.com/downloads2/send/5-papers/278-study-about-the-rapid-manufacturing-of-complex-parts-of-stainless-steel-and-titanium*

[5] Gebhardt, A.: For the Third Industrial Revolution – Trend Report Technologies. *Kunststoffe International* 10/2009, pp. 88-93

[6] Fateri, M.: *Selective Laser Melting of Glass Powders*, Dr. Hut Verlag, Munich, 2018

[7] Gebhardt, A.: Vision Rapid Prototyping – Generative Manufacturing of Ceramic Parts – A Survey Proceedings of the German Ceramic Society, (Deutsche Keramische Gesellschaft e.V.), DKG 83 (2006) No. 13

[8] Klosterman, D. A., Chartoff, R. P., Osborne, N. R, Lightman, A., Han, G., Graves, G., Bezeredi, A., Rodrigues, S.: Development of a Curved Layer LOM Process for Monolithic Ceramics and Ceramic Matrix Composites. Rapid Prototype Development Laboratory University of Dayton, OH, USA – *Journal of Rapid Prototyping*, Vol. 5, No. 2, 1999, pp. 61–71

[9] Mavroidis, C., DeLaurentis, K. J., Won, J., Alam, M.: Fabrication of Non-Assembly Mechanisms and Robotic Systems Using Rapid Prototyping. *Transactions of the ASME, Journal of Mechanical Design*, Vol. 123, 12/2001, pp. 516–524

[10] N.N.: VDI 3405 Blatt 3 Additive Fertigungsverfahren – Konstruktionsempfehlungen für die Bauteilfertigung mit Laser-Sintern und Laser-Strahlschmelzen – English: Additive manufacturing processes, rapid manufacturing – Design rules for part production using laser sintering and laser melting. Published: 2015-12. VDI Gesellschaft Produktionstechnik (ADB), Postfach 10 11 39, D-40002 Düsseldorf/Germany, 2015

[11] Pfefferkorn, F.: EOS Polymer Laser-Sintern – Möglichkeiten und Einschränkungen bei der Bauteilauslegung. Workshop: Laserbearbeitung von Kunststoffen, Bayerisches Laserzentrum, Erlangen, July 5–6, 2011

[12] König, P., Barczok, A.: Ideen materialisieren. Webdienste fertigen Objekte nach Ihren 3D-Entwürfen. *c't magazin für computer technik*, Issue 15, 2011, pp 84–94; Heise Zeitschriften Verlag GmbH & Co. KG, Hannover, Germany

7 Glossary

List of frequently used terms and their abbreviations that are often used in literature but not always clearly explained there.

Abbreviation	Definition	Explanation
3DP	Three-Dimensional Printing	Rapid prototyping process. Layer formation by injecting liquid into powder. Trademark Massachusetts Institute of Technology
3DP	Three-Dimensional Printing	Generally accepted generic term for all layer manufacturing methods. Replaces the standardized term "additive manufacturing"
ACES	Accurate Clear Epoxy Solids	Stereolithography build style (3D Systems)
AF	Additive Fabrication	Production by adding volume elements
AF	Anatomic Facsimiles	Medical anatomical (surgery) models
AFM	Anatomic Models	Medical anatomical (surgery) models
AIM	ACES Injection Molding	Injection molding tool made from stereo-lithography resin using the ACES build style
ALM	Additive Layer Manufacturing	Production by adding volume element (layers). Proprietary term originally from Boeing, commonly used in aviation industry
AM	Additive Manufacturing	Production by adding volume elements (layers)
AMF	Additive Manufacturing File Format	Extended STL format, supporting coloring, multi material and graded materials, textures, or microstructures. Established by the ASTM (founded as the American Society for Testing and Materials)
BASS	Break Away Support System	Support structure for the FLM process, which can be removed manually (Stratasys)
BIS	Beam Interference Solidification	Rapid prototyping process. Polymerization area defined by two intersecting laser beams
	Bridge tooling	Simple tools for "bridging" the gap between prototype tooling and serial tools
	Buy-to-fly ratio	Ratio of the volume of material to be paid for and the material forming the part: a measure of the waste in a manufacturing process

Abbreviation	Definition	Explanation
CAD	Computer-Aided Design	Computer-aided design, particularly in the sense of geometric construction
CAE	Computer-Aided Engineering	Engineering design using CAD
CAL	Computer-Aided Logistics	Computer-aided material storage, distribution, and supply
CAM	Computer-Aided Manufacturing	Computer-aided manufacturing
CAMOD	Computer-Aided Modeling Devices	Software and hardware elements for the computer aided production of models
CAP	Computer-Aided Production	Computer-aided production
CAQ	Computer-Aided Quality Assurance	Computer-aided quality assurance
CAS	Chemical Abstracts Service	US-American nomenclature and institution for the identification of chemical substances
CAS	Computer-Aided Styling	Computer-aided shaping (styling)
CAT	Computer-Aided Testing	Computer-aided measurement and testing
CAx	Computer-Aided ... Computer-Assisted ...	Abbreviation of any kind of computer-aided processes
CIM	Computer-Integrated Manufacturing	Production based on a closed CAD-CAM chain. Equivalent to ICAM
CIM	Computer-Aided Manufacturing	Production on the basis of a closed CAD-CAM chain
CJP	ColorJet Printing Technology	3D Systems designation of the colored powder binder process (ex. Z-Corp)
CLIP	Continuous Liquid Interface Production	Continuously operating polymerization printer by Carbon 3D
CMB	Controlled Metal Build Up	Rapid prototyping process. Laser generation and subsequent contour milling of metal (Fraunhofer-IPT)
CP	Centrum für Prototypenbau GmbH	Rapid prototyping service provider located in Erkelenz (Germany)
CPDM	CIMATRON Product Data Management	PDM System of the CIMATRON company
CS ...	Computer Supported ...	Computer-aided ... process
CSG	Constructive Solid Geometry	Description of a complex body by combining simple elements by Boolean operations
CT	Computed Tomography	X-ray based scanning process, used particularly in medicine
DCM	Direct Composite Manufacturing	Direct fabrication of components made of composite material. Predominantly used by OptoForm for the M3D Process
DLP	Digital light processing	Similar process to stereolithography which uses a lamp and mask
DMLM	Direct Metal Laser Melting	Process like DMLS, where the metal is completely molten. Predominantly used by GE
DMLS	Direct Metal Laser Sintering	Proprietary designation of the metal AM process of EOS

Abbreviation	Definition	Explanation
DMP	Direct Metal Printing	Proprietary designation of the metal AM process of 3D Systems. Originally: Metal AM process of the Belgian company LayerWise (acquired 2014 by 3D Systems)
DMU	Digital Mock Up	Digital prototype, mostly in the sense of animatable files in virtual reality applications
DoB	Drop-on-Bed	3D Printing, Powder Binder process of MIT, equivalent to DoP
DoD	Drop-on-Drop	3D printing process, designation predominantly by the Sanders company
DoP	Drop-on-Powder	3D Printing, Powder Binder process by MIT, equivalent to DoB
DT	Direct Tooling	Additive processes for the direct production of metal tool inserts, molds, and dies
DTM	Desktop Manufacturing	Process for the fast production of three-dimensional physical models "on the table or desk"
DTM	DTM Corp., Austin, TX, USA	Manufacturer of laser sintering machines; now 3D Systems
DXF	Drawing Exchange Format	Data format for storing CAD drawings
EBFF	Electron Beam Free Form Fabrication	AM process for the production of metal parts by means of an electron beam
EBM	Electron Beam Melting	AM process for the production of metal parts by means of an electron beam predominantly used by the ARCAM company (now GE)
EDM	Electronic Data Management / Engineering Data Management	Program systems for managing large files that can be simultaneously processed by multiple authorized people
EDM	Electrical Discharge Machining	Shaping by electrical discharge
EDM	Electronic Document Management	Electronic archiving of documents
EOS	Electrical Optical Systems GmbH, Planegg/Munich, Germany	Manufacturer of sintering machinery
ERM	Enterprise Resource Management	System and software program for the management and planning of logistics supply chains
ERP	Enterprise Resource Planning	System and software programs for the management and planning of logistics supply chains
	Fabber	Short form of Fabricator
	Fabricator (Fabrikator)	Additive manufacturing machine for the direct production of components with product character
FDM	Fused Deposition Modeling	FLM process of the Stratasys company. Proprietary, non-common term for the FLM extrusion process
FEM	Finite Element Method	Numerical calculation and simulation method
FFF	Fused Filament Fabrication	Alternative expression for Fused Deposition Modeling (FDM / FLM)

Abbreviation	Definition	Explanation
FFF	Fast Freeform Fabrication	Process for the rapid production of three-dimensional physical models
FFM	Fused Filament Manufacturing	Alternative expression for Fused Deposition Modeling (FDM / FLM)
	Finishing	AM process: independent treatment of AM parts, especially to achieve better surface qualities, for example by grinding, painting, or polishing.
FGM	Functional Graded Manufacturing	AM of parts with functionalized gradient materials
FLM	Fused Layer Modeling Fused Layer Manufacturing	Alternative expression for Fused Deposition Modeling (FDM) and filament deposition
FM	Facsimile Models	1:1 scale 3D models
	Front end	Technologically the front end of the AM process chain. The front end defines the chain elements between the definition of the geo-metric data set and its transfer to the control unit of the machine.In general: The first ele-ments of a process chain
GIS	Geographic Information System	Information systems for recording, processing, organization, analysis, and presentation of spatial data
	Grower	Designation for AM machinery, mainly used by Solidscape for the T66 machine
HIS	Holographic Interference Solidification	Rapid prototyping process. Polymerization by projecting (3D) holographic images onto photosensitive materials
HPGL	Hewlett Packard Graphic Language	Plotter interface
HSC	High Speed Cutting	High speed milling
HSPC	High Speed Precision Cutting	Proprietary term used by the Kern Microtechnik company for its HSC machines
ICAM	Integrated Computer Aided Manu-facturing	Production based on a closed-chain CAD-CAM; equivalent to CIM
IGES	Initial Graphics Exchange Specification	File format for the exchange of neutral geometry data between CAD systems
IwF	Institute for Toolless Production	Additive manufacturing research provider located in Aachen (Germany)
	Indirect rapid prototyping process	Additive processes that do not directly work digitally, but uses AM parts, mainly for counter casting
CMM	Coordinate Measuring Machine	Machine for obtaining the 3D coordinates of a physical sample
LAM	Laser Additive Manufacturing	Topic for SLM and EBM
LBM	Laser Beam Melting	Equivalent to SLM; used in contrast to EBM
LCVD	Laser Chemical Vapor Deposition	Additive process. Laser-assisted deposition of material from the gas phase

Abbreviation	Definition	Explanation
LENS	Laser Engineered Net Shaping	Proprietary term for metal process of the OPTOMEC Company. Equivalent to LMD
LLM	Layer Laminate Manufacturing	AM process. Contouring layers from sheets or foils by laser, knife, or cutter. Part building by bonding of subsequent layers
LM	Laminate Manufacturing	Equivalent to LLM
LMC	Layer Milling Center	Proprietary term for a LM machine of the Zimmermann company
LMD	Laser Metal Deposition	Synonymous for LAM, particularly for powder-based processes
LMF	Laser Metal Fusion	Proprietary term for the SLM process of the company Trumpf / Sisma
LMI	Laser Melting Innovations	Manufacturer of SLM machines (Aachen Germany)
LMP	Layer Milling Process	AM process by layer milling of the Zimmermann/Pauser company
LMPM	Low Melting Point Metal	Metal alloy with a defined and low-melting point
LMS	Laser Model System	Proprietary Stereolithography process of the Fockele & Schwarze company (now: Realizer)
LMT	Layer Manufacturing Technologies & Techniques	General (generic) term for layer manufacturing processes
LOM	Laminated Object Manufacturing	LLM process of the Helisys company
L-P-BF	Laser-Powder-bed Fusion	Processes with single-component metal powders for the additive manufacturing production of components with the product character (final products)
LS	Laser Sintering	Additive manufacturing process: layer formation by local melting and subsequent solidification of powder-like materials
LSM	Laser Surface Melting	Metal laser-sintering process of the Realizer company (ex. Fockele & Schwarze)
M3D	Maskless Mesoscale Material Deposition	Additive manufacturing process of the OPTOMEC company based on aerosol printing
MEMS	Micro Electromechanical Systems	Miniaturized functional system made from mechanical, electro-mechanical, or optical systems
MIM	Metal Injection Molding	Injection-molding process based on plastifiable metal mixtures from metal powders and binders
MIM/ MAM/ MDM	Material Incress Manufacturing/ Material Addition Manufacturing/ Material Deposition Manufacturing	Manufacturing of parts by successively adding material (particularly by layers)
MJM	Multi-Jet Modeling	PJM process of the 3D Systems company

Abbreviation	Definition	Explanation
MJS	Multiphase Jet Solidification	PJM process of the ITP company (no longer in business)
	Modding	Upgrading of products mostly by fitting accessories or ornamentation to otherwise unchanged products. Equivalent to tuning, styling, pimping
	Model	"Show and tell" component for the visualization of a product. Sometimes with functionalities
MPA	Metal Powder Application	Proprietary thermal spraying process from the Hermle company. Integrated into the Hermle five-axis machining center
MRI	*Magnetic Resonance Imaging*	Medical imaging process, particularly for the examination of soft tissue
OEM	Original Equipment Manufacturer	Supplier for products that the original manufacturer sells under its own name
	Pimp	Especially externally spectacular treatment of a product that is otherwise unchanged, especially in its function. See also modding
PDM	Product Data Management	Electronic system for product data management
PET	Positron Emission Tomography	Imaging process predominantly used in medicine
PJM	PolyJet Modeling	Layer-by-layer additive technique in which liquid photopolymer resins (polymers with photo-activators) are deposited line-by-line and immedi-ately harden on exposure to UV radiation
PLA	Polylactide	Bio-plastic material
PPS	Production Planning System	Production planning and control system
	Prototyper	Additive manufacturing machine for the pro-duction of prototypes, patterns, and dummies
RCT	Rapid Casting Technology	Term for additively supported casting processes. Especially used by Prometal / ExOne to manufacture sand cores and shapes
RIM	Reaction Injection Molding	Injection-molding processes on the basis of plasticizable chemically reactive metal–plastic mixtures
RM	Rapid Modeling	Process for the rapid production of models
RP	Rapid Prototyping	Technology that deals with processes and methods for the production of layered models directly from 3D CAD data
RP	Reinforced Plastics	Predominantly fiber-reinforced plastics (as opposed to non-reinforced plastics)
RP&M	Rapid Prototyping & Manufacturing	General term for processes that rapidly make prototypes and products. Mainly additively layer by layer

Abbreviation	Definition	Explanation
RPD	Rapid Product Development	Rapid product development
RPro	Rapid Production	Rapid manufacturing or rapid production
RPT	Rapid Prototyping Techniques/ Technologies	Processes and methods for rapidly making prototypes. Mainly additively layer by layer
RT	Rapid Tooling	Additive processes for making molds, dies, and tools
RTV	Room-Temperature Vulcanization	Follow-up process, also known as vacuum casting or silicone casting
SAHP	Selective Adhesive and Hot Press Process	LLM process of the KIRA company
SDU	Shell Design Unit	Hardware and software system of the Soligen company for the design and manufacture of ceramic shells or molds (negatives)
SE	Simultaneous Engineering	Methodic approach for parallel work of several people or teams on the same (development) task
SET	Standard Déchange et de Transfer	File format for neutral exchange of geometric data sets between CAD systems
SFF	Solid Freeform Fabrication	Additive process for making physical parts (solids). Equivalent of SFM
SFM	Solid Freeform Manufacturing	Equivalent of SFF
SFP	Solid Foil Polymerization	Additive manufacturing process. Parts are made from contoured layers joined by polymerization
SGC	Solid Ground Curing	Stereolithography process of the Cubital company
SL	Stereolithography	Additive manufacturing process. Parts are made from layers build by local solidification of photosensitive resins (photopolymerization)
SLA	Stereolithography Apparatus	Stereolithography system of the 3D Systems company
SLM	Selective Laser Melting	L-P-BF process of the Realizer company
SLPR	Selective Laser Powder Remelting	Additive manufacturing process. Parts are made by local melting and re-solidification of one component metal powder (Fraunhofer–ILT)
SLS	Selective Laser Sintering	Additive manufacturing process. Parts are made by local melting and subsequent solidification of thermoplastic powder, Equivalent of LS
SOUP	Solid Object Ultraviolet Laser Plotter	Additive manufacturing process and system based on stereolithography (CMET company)
SPECT	Single Photon Emission Computed Tomography	Imaging process used predominantly in medicine
SPF	Super Plast Forming	Parts are made by "inflation" of sandwich structures

Abbreviation	Definition	Explanation
STAR-Weave	Staggered Alternated Retracted Hatch	Stereolithography build style (3D Systems).
STEP	Standard of Exchange of Product Model Data	File format for the neutral exchange of the entire product data between CAx systems
STL	Stereolithography Language	Interface format for the exchange of geometry data between CAD systems and additive manufacturing machines. Originally: "standard transformation language". Initially developed for simple shading of 3D CAD structures
TCT	Time Compressing Technologies	Collective name for all processes capable of shortening the product development time (process)
THESA	Thermoelastic Stress Analysis	Thermoelastic tension analysis. Process for the experimental verification of component stress by measuring thermal effects
TI	Tailored Implants	Individually "tailored" implants
TP	Thermal Polymerization	Polymerization by applying heat
UV	Ultraviolet	Electromagnetic radiation with a wavelength from 10 to 400 nm
VDAFS	German Association of the Automotive Industry, Surface Interface	CAD data exchange format for the transfer of surface models from one CAD system to another
VDAIS	German Association of the Automotive Industry, IGES Interface	CAD data exchange format for the transfer of CAD models from one CAD system to another. Represented by a subset of the elements defined by IGES
VR	Virtual Reality	Realistic simulation of components, assemblies, or entire products on the computer, usually combined with real-time animation. Input and output through data gloves, 3D projection, or the like

Index

Quick, Individual, Decentralized

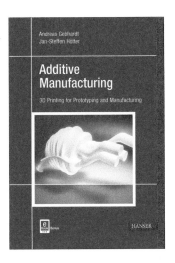

Gebhardt, Hötter
Additive Manufacturing
3D Printing for Prototyping and Manufacturing
611 pages. E-Book Bonus. In full color
$ 199.99. ISBN 978-1-56990-582-1
Also available separately as an E-Book

The use of additive manufacturing for the direct production of finished products is becoming increasingly important.

Oriented towards the practitioner, in this book the basics of additive manufacturing are presented and the properties and special aspects of industrially available machines are discussed. From the generation of data to the forming method, the complete process chain is shown in a practical light.

Applications for the production of models and prototypes (rapid prototyping), tools, tool inserts, and forms (rapid tooling) as well as end products (rapid manufacturing) are covered in detailed chapters with examples. Questions of efficiency are discussed from a strategic point of view, and also from an operational perspective.

Crossing the border between proto-typing and functional components

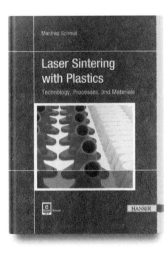

Laser Sintering (LS) with plastics is one of the most promising additive manufacturing technologies: it is currently regarded as the process most likely in the future to permanently cross the border between prototyping and the production of functional parts. This step is challenging because it means that the technology must meet certain requirements that are also valid for traditional and established production processes. Only by succeeding at this step can a wide industry acceptance of LS be expected in the future.